VEHICLE DYNAMICS
AND DAMPING

VEHICLE DYNAMICS AND DAMPING

First revised edition

JAN ZUIJDIJK

authorHOUSE®

AuthorHouse™
1663 Liberty Drive
Bloomington, IN 47403
www.authorhouse.com
Phone: 1-800-839-8640

© 2013 by Jan Zuijdijk. All rights reserved.

No part of this book may be reproduced, stored in a retrieval system, or transmitted by any means without the written permission of the author.

Published by AuthorHouse   03/20/2013

ISBN: 978-1-4772-4736-5 (sc)
ISBN: 978-1-4772-4737-2 (hc)
ISBN: 978-1-4772-4738-9 (e)

Any people depicted in stock imagery provided by Thinkstock are models, and such images are being used for illustrative purposes only.
Certain stock imagery © Thinkstock.

This book is printed on acid-free paper.

Because of the dynamic nature of the Internet, any web addresses or links contained in this book may have changed since publication and may no longer be valid. The views expressed in this work are solely those of the author and do not necessarily reflect the views of the publisher, and the publisher hereby disclaims any responsibility for them.

CONTENTS

Foreword vii
Acknowledgements ix
Statement of non-liability xi
Preface xiii

Chapter 1:   Learning the science of damping 1
Chapter 2:   The art of damping design 4
Chapter 3:   Damping characteristics 16
Chapter 4:   Shims 21
Chapter 5:   Interpretation of damper graphs 24
Chapter 6:   Springs 30
Chapter 7:   Dynamometer 33
Chapter 8:   Twin-tube damper designs 51
Chapter 9:   Development Of The Twin-Tube Hydraulic
             Damper With External Adjustment Knobs 58
Chapter 10: Flow Damper Design 63
Chapter 11: Twin tube damper 65
Chapter 12: McPherson strut design and application for
             sport and racing car 72
Chapter 13: The Damping Function Of The Twin Tube
             Strut/Damper 82
Chapter 14: McPherson Struts For Race Applications 84
Chapter 15: Single/Mono tube damper design and operation 87
Chapter 16: Mono Tube Damper With External Gas/Fliuid
             Reservoir 95
Chapter 17: Low Pressure Twin Tube Gas Damper 99

Chapter 18: Function Of The External Gas
And Fluid Reservoir 105
Chapter 19: Experience with compression damping 112
Chapter 20: Development of the self-leveling suspension for
the Porsche 911 and Ferrari 365 2+2 125
Chapter 21: Front Wheel Driven Race Cars 132
Chapter 22: Development Of The Dashboard Adjustable
Damper 136
Chapter 23: Effects of damping on cornering 142
Chapter 24: Forward Traction 149
Chapter 25: Damping and lateral grip.. 155
Chapter 26: Compression damping of the four way adjustable
damper 161
Chapter 27: Relation between a damper running in a dyno
and performing in the suspension............................. 173
Chapter 28: Working at the racetrack 177
Chapter 29: Adjusting The Two Way Adjustable Damper........... 186
Chapter 30: Single Tube Damper With External Reservoir.......... 193
Chapter 31: Using Bumprubbers.. 200
Chapter 32: Adjusting The Rebound Damping Force
Of The Threeway Adjustable Damper 202
Chapter 33: Adjusting The Fourway Adjustable Damper............. 204
Chapter 34: Force Velocity Graph Versus
Force Displacement Graph.. 209
Chapter 35: Back ground Jan Zuijdijk.. 211

FOREWORD
by Ron Fellows.

I am honoured and feel very privileged that Jan has asked me to write a few words for his book.

I first met Jan Zuijdijk back in 1995. It was during my first year racing for the Chevrolet Camaro factory team when we were competing in the SCCA Trans Am Series. Our American Equipment Racing team had started developing a new chassis for that season and our team manager/engineer Will Moody, invited Jan and his (new) JRZ damper team to one of the test sessions.

I took an instant liking to Jan. He was friendly, always smiling, soft spoken, with an incredible knowledge of dampers and their inter-action with the car and tire. And back then, the list of drivers that he had worked with was a who's who of motor racing!

The JRZ dampers were an immediate asset on our Camaro. With Jan's assistance, our Camaro team dominated in the win column over the next two seasons. And with the first chassis that I raced, we managed to win six of twelve races!

The automotive community should be very pleased that Jan has taken the time to write a book about his passion. His years of experience, knowledge of dampers and suspension tuning is likely second to none. This book will provide a great knowledge base for the next generation. And I can't wait to get my copy!

Ron Fellows

Toronto, Canada

Acknowledgements

I like to express my thanks to the following people for their support with this book

Ron Fellows, for writing such a nice and sincere foreword.

Erik Ras, manager C.O.O of JRZ Suspension Engineering for providing me with drawings and sketches.

Marcel de Klein, JRZ, for 3D drawings of special components.

Tom Milner, Prototype Racing Group for the race car pictures.

My Wife Jannie, spending my days in my office writing.

Statement of Non-liability

Although the safety of motorsport is improved contineously, there are aspects of the sport which do not take away the fact that a certain degree of danger is present for participants in motorsport events.

Competition requires participants in motorsports events to stretch the limits of performance.

While this book is aiming to improve the competitive level of handling and performance of race and sports cars, all participants in competitive events should be aware that applying any of the procedures, Idea's or advice written in this book, they do so at their own free will making their own decision.

I disclaim any responsibility for their actions or accidents which might occur.

Jan Zuijdijk.

Preface

The content of this book is for Engineers, racers, students or anyone who would like to know more about the working of dampers, adjustments, the effects of damping control on the setup of a race car and vehicle dynamics.

There are many miss understandings about the hydraulic processes taking place inside the damper.

Thts book offers an insight on how different damping systems work en how they affect the performance and handling of race and high performance cars from a suspension engineers point of view.

Development is learning from earlier experiences; development of a damping system, parts and components is a never ending process, building on past experiences we are still finding new ways to improve.

Looking back at the time since we got involved in suspension design for racing cars, the science of suspension design has made great strides forward.

What we have achieved in product development today is a world of difference compared to the early days of suspension development, when we got involved in suspension and damper design.

The state of the art of current design is a four way adjustable damper which is an instrument, build with many little parts, each performing their specific role in the complex world of vehicle dynamics and damping.

In addition to the hydraulic damping performance, I introduced the variable lift, produced by gas pressure which in combination with the damping performance is very beneficial for the handling of a car and to show how far ahead of its time it was, it is presently being copied more than once.

We at JRZ have developed an unique spring/damping system in which the damper carries part of the weight of the car with variable lift in combination with a fourway adjustable damping system reducing the importance of the steel spring.

The benefits of this combination spring/damper system proves itself on a daily basis in race series in many parts of the world, our customers find improved lap times and better handling just switching to a set of JRZ Suspension components.

We arrived at this level of understanding the role of damping in vehicle dynamics through constantly evaluating and learning from our mistakes, expanding our knowledge in the field of damper and suspension design and combining past experience with new developments in damping, keeping in mind that the damper is only one factor in the complexity of vehicle dynamics of a car in motion.

I have not expanded on future developments in this field, but I am sure that as long as there are people building and racing cars, new ways will be found to make the cars go safer and faster at speed.

In this time and age in which motor racing seems to be taken over by electronic controls, one fact remains essential, no matter how advanced the electronic equipment is, a race car still runs on four tires with a contact patch no larger than a few square inches, keeping those small pieces of rubber firmly in contact with the road surface under all circumstances is our objective and our specialty.

If anyone can find something in this book that helps them to make better use of the damper / suspension, increasing their performance I considered it a success.

Jan Zuijdijk, author.

01

LEARNING THE SCIENCE OF DAMPING

My career started at Koni as damper/suspension specialist in the function of R and D engineer.

Working in the Koni R & d department I quickly learned that there is a lot more to the design and application of dampers than just a tubular part working silently under a car.

The main Koni business during the year of 1960 was developing dampers for road car, bus and truck applications in addition to car heaters, car and truck lifts etc.

Koni was becoming a major player in the railway business, developing very heavy dampers for railway all over the world and later on in the development of the high speed train TGV in France.

My role was to be in development of dampers for road cars extending itself into development of race damper applications, from Proto type, Sports cars to Formula One, Ferrari was one of the early customers.

I could have never guessed that my R and D job at Koni would open the door to my involvement with all kinds of racing cars, in particular Formula One and sports cars and racing cars, travelling all over the world, working and living in the USA for 20 years, meeting and working with famous race teams, drivers, industrial engineers and business executives.

The racing community had just evolved from using friction and rotary dampers changing to telescopic type dampers and because the Koni dampers were adjustable in rebound, in order to perform the adjustment they had to be removed from the car but they were a couple of steps ahead of the competition.

Suspension and damping were not a great concern to drivers, race engineers and technicians at that point in time, race engineering if it can be called that way, was done by the team owner or the engineer who designed the car, even laywers could be race engineers so I got in on the ground floor so to speak.

Prince Bernhard of the Netherlands was friends with Arie and Kor de Koning, the two brothers who founded koni right after WW2 but He was also a Ferrari enthusiast, He used to visit the Koni factory once in a while and bring His new Ferrari to be equipped with a new set of dampers taking the car back to Maranello impressing Mr Enzo Ferrari with the improved ride and handling of the car so the story goes and as a result, Koni was invited to develop new suspension products for the Ferrari cars.

Soon thereafter Koni dampers were original equipment on the Ferrari production cars. The Ferrari formula One cars where the first racing cars equipped with telescopic Koni dampers, soon to be followed by the Porsche Formula One factory team.

The formula one cars of this era were still painted in the national colours, Ferrari Red, Porsche Silver, British Green etc.

The Koni racing service, at that point, consisted of not much more than attending a race event with some parts and tools in the back of a passenger car.

The dampers could be checked with the help of a "portable" little dyno but this was a task to be avoided because the lack of electrical power in most race paddocks.

The dyno was little and fitted in the back of a station wagon but it was made of heavy steel plates and so heavy that the front wheels of the stationwagon almost lifted off the road when it was loaded into the car.

In the early days of racing service Koni used a service van, it was a used Renault delivery van, overloaded with too much equipment and with very little room to work inside, in order to operate the little dyno, we had to slide it on rails to the side door where it could be operated by a technician standing outside the van.

The paddock facilities at most race tracks were limited, the race cars, after practice, Formula One and sports car teams were using garages in different towns in the vicinity of the track, especially in Le Mans and Franchorchamps, we had to travel around and visit the teams in their different locations to service, test, adjust and change damper characteristics to the drivers wishes.

The Formula One cars driven by chief mechanics over the open road from the race track to the garage where they worked on the cars was a special attraction to the public.

The dyno ran on 220 volt alternating current, a voltage which was not to be found in many places in Belgium and France, in addition to this the wall plugs were different in every other village, so we spend more time changing wall-plugs than actually working on dampers.

It was a more exciting time because the teams were not confined to their garages and motorhomes like it is today but we enjoyed the open and friendly atmosphere, it was more a kind of club atmosphere where you could meet drivers, team owners and every one else you worked with the garages and in the hotel having dinner or a drink at the bar.

I am proud to say that thanks to my early Job at Koni and later on in my own business, stubbornly moving forward with new insights and developments, always questioning traditional thinking, sometimes against all odds, challenging existing levels of design, have made a contribution to achieve a higher level of suspension design and performance for all kinds of race and sports car applications.

02

THE ART OF DAMPING DESIGN

Dampers, In the USA referred to as "shock absorbers" or "shocks ", is a popular name for a technical devise performing more important and diverse functions than the name indicates.

The importance of the controlling function in what is called "Vehicle Dynamics" is not always properly recognized.

A better understanding how the damper works and what it can do or not do will help people make better decisions when tuning the suspension of a race car, obeying laws of physics instead of using the trial and error method.

There is a much better understanding of the science of damping now as a result of research and development leading to the design of today's multi-functional device which is an important part of modern suspension.

Shock absorbers do not absorb shocks, the name is incorrect, and we prefer calling this device a damper.

The basic purpose of the damper in a suspension is converting energy of motion or kinetic energy generated by the moving mass of a vehicle stored in the compressed spring, into heat or thermal energy

The thermal energy is than dissipated into the airstream through the metal damper body which is acting as a heat exchanger.

The compression of the suspension is controlled by damping forces, slowing down suspension travel in compression, preventing the springs from absorbing more kinetic energy than necessary than when this process would be undampened.

The damper slows down, controls, the motion of the suspension in both rebound and compression.

Every aspect of hydraulic damping is subjected to constant research and development, being improved and tested in the lab and in the field.

Hydraulic damping is still a mysterious component in the complexity of suspension tuning, a basic understanding of damping, how it functions in the suspension and the effects of damping on handling and vehicle dynamics will be very helpful for anybody engaged in car/suspension design and race car engineering.

Designing a damping characteristic is certainly not a hit or miss proposition or a black art as popular believe has it, the processes are based on sound engineering practices, governed by laws of physics, one cannot change laws of physics.

In addition to understanding the science of damping, damper engineers also need know what part of the vehicle dynamics we want to control with damping.

First: a damper is subject to hydraulic laws.

The compression damping forces are generated according to the principle that "no two solid or liquid bodies can take up the same space at the same time"

A piston rod inserted in a cylinder filled with fluid will displace an amount of fluid equal to the volume of the piston rod, the amount of fluid is forced out of the cylinder and can only flow through the valving system into the fluid reservoir generating damping forces.

By the twintube damper the fluid reservoir is the cylindrical space between the outer cylinderwall and the inner diameter of the damper body.

In the mono-tube gas/hydraulic damper the bottom part of the working cylinder functions as reservoir with a floating divider piston, keeping gas volume and fluid volume separated from each other.

The single tube gas/hydraulic damper can be equipped with an external reservoir absorbing the amount of fluid displaced out of the cylinder, saving packaging lengt which is important when a short damper is required.

Leading the fluid through valves and orifices we create damping forces.

The amount of damping is determined by the number of orifices and spring loaded valves.

During the rebound stroke when the rod is pulled out of the cylinder, pressurized fluid trapped between the piston and guide flows through the Piston valving system to the lower part of the cylinder while at the same time the amount of fluid which was displaced out of the cylinder during the compression stroke is flowing back into the cylinder through the non return valve in the footvalve taking the vacant place of the withdrawing piston rod.

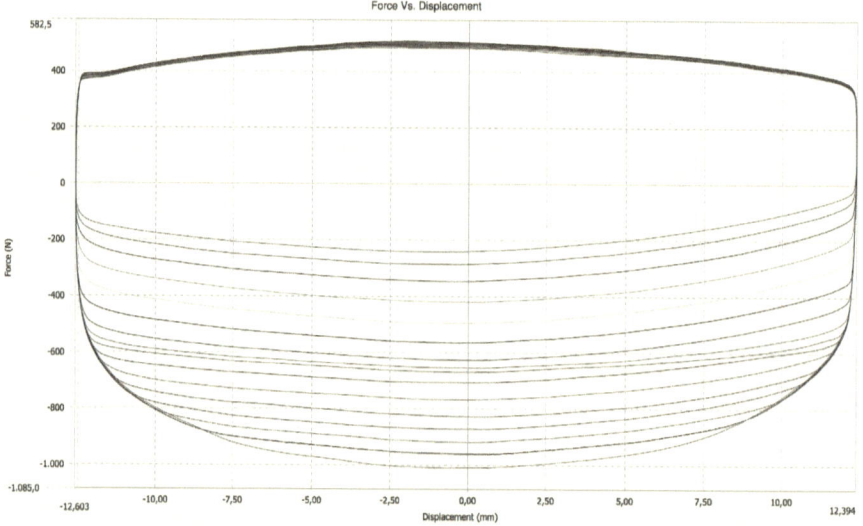

VEHICLE DYNAMICS AND DAMPING

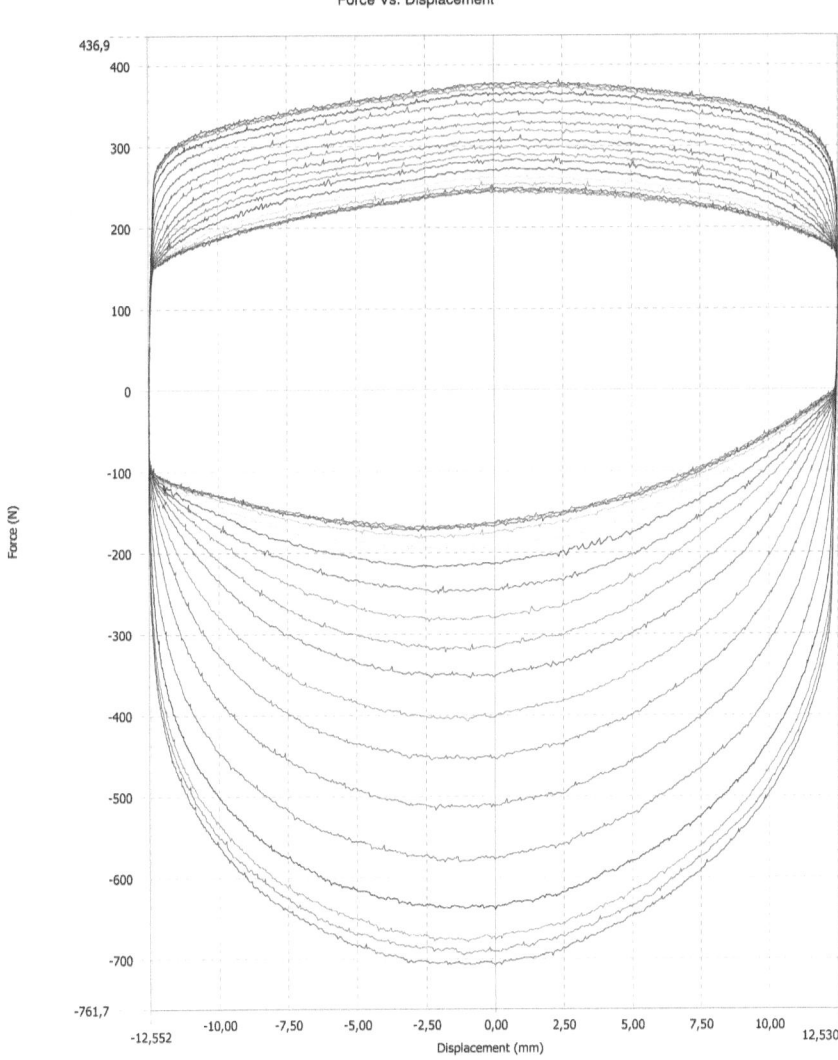

Rebound adjustments
Force vs. Displacement

No two solid or fluid bodies can take the same space at the same time.

When a piston rod is inserted into a cylinder filled with fluid, an amount of fluid equal to the piston rod volume is forced out of the cylinder.

The amount of fluid which is displaced in compression, depends on the piston rod diameter, a 22 millimetre diameter piston rod displaces 3,8 cubic centimetres or 0.589 cubic inch each centimetre stroke,

A 12 millimetre diameter piston rod displaces only 1 cubic centimetre or 0.155 cubic inch fluid per centimetre stroke, so a large displacement piston rod generates more damping force at low piston velocities in compression.

The second principle which applies to hydraulic damping is that a pressure exerted on fluid in a closed vessel is transmitted throughout the entire system with the same pressure according to Pascal's law.

Following Pascal's law we can use a remote reservoir with the hydraulic controls (adjustable valving) built-in the valve housing of the remote reservoir.

The pressure, transported through a high pressure line into the valving system of the remote reservoir is the same everywhere in the system, thus also in the remote reservoir.

Hydraulic fluid is a wonderful medium to transmit pressures in a hydraulic system.

Gas pressure

Using gas as a medium providing lift is based on the same principle that pressure exerted on a gas volume is transmitted throughout the system with equal pressure.

Calculating the pressure increase or decrease in a vessel with a changing volume, we can use boyle's law which says" **pressure x volume is constant.**

It means that when the pressure is increased, the gas volume is reduced, following this law we can calculate the increase in gas pressure in the gas/oil reservoir when the damper is compressed from extended length to minimum length.

How this principle can be applied in damper design is explained with the figures on the next page.

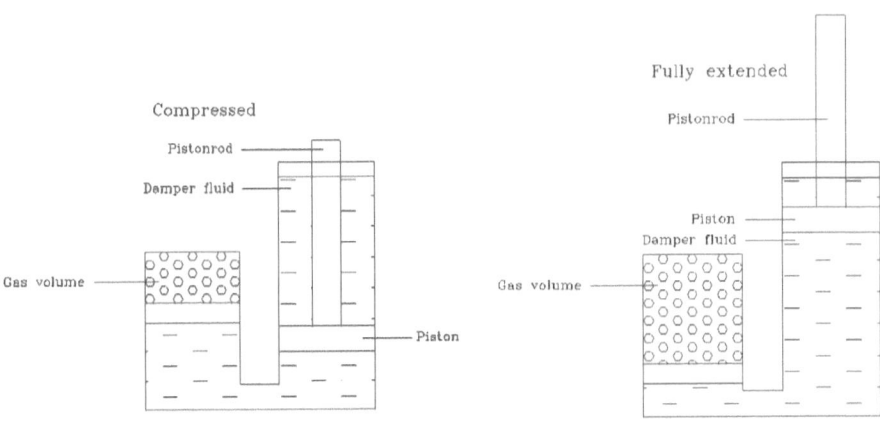

Example.

In a damper with a fully extended piston rod, the gas volume in the reservoir is 160 cubic centimetre

The gas pressure is 20 bar, the constant factor to calculate the pressure is 20 X 160 = 3200.

With the piston rod fully compressed with a 10 Centimeter stroke the gas volume is reduced from 160-38 cubic centimetre to 122 cubic centimetre.

The gas pressure is than 3200 divided by 122 = 26 bar.

The increase in gas force per centimetre stroke is than 0,6 bar so the spring rate in this system originating from the gas pressure is 0,6 kg per Centimetre which is very little.

If the pressure increase is to much, the gas volume in the reservoir needs to be increased.

This can be done using a larger diameter reservoir or the length can be increased.

A larger diameter reservoir has the advantage that the travel ratio piston rod/divider piston in the reservoir is better.

In this design the divider piston will use less travel.

By changing the gas volume up or down, the lifting force can be increased or decreased which means that a certain variable low spring rate can be built-in to the system combining the damping function with a load carrying function.

Changing the gas pressure, the lifting force can be varied from a low of 20 kg (44 pounds) to a high of 95 kg (210 pounds) per corner of a car.

The gas pressure exerts a force on the cross section of the piston rod so the lifting force is dictated by the gas pressure multiplied bij de cross section of the piston rod in centimeters square.

We have used this feature very successful in many applications; the gas lift provides traction, grip, transmitting engine power better to the pavement.

The gas lifting force works two ways; it supports the spring to carry the weight of the car and keeps the tire contact patch in firm contact with the road surface.

The lift produced by the gas pressure carries part of the weight of the car so that the steel spring is compressed less than would be without the gas lifting force.

Newton's law.

When tuning dampers in the suspension system of a car at the race track, we have to comply with newton's laws of motion, acceleration, gravity etc.

Newton's law says that an object (car) brought in motion it wants to continue in the same direction with the same speed and force, Action = Reaction. (Inertia)

Steering a car around a corner is contrary to what the car wants to do according to Mr Newton's law, the car wants to continue in a straight line with the same speed and force with which it is brought into motion.

Cornering is against the principles of this law of physics but Increasing the grip level and traction of a car can help the car run the corner with greater speed and that is what we can do with a better damper design, so

VEHICLE DYNAMICS AND DAMPING

we are always trying to find ways to stretch the limits of lateral grip and adhesion.

Laws of physics play an important role in damper design, it is impossible to ignore them. mr Newton discovered gravity, a force which we have to adhear to in many different ways.

Vehicle dynamics is a complexity of forces acting upon a car in motion, with a good designed damping characteristic, understanding the correct use of the damper adjustments we can control those forces helping the suspension to transmit engine power, steering input and braking power optimal to the pavement

Damping forces in rebound and compression are generated by the flow of fluid through a combination of orifices, valves, piston velocity and the amount of fluid displaced per centimetre, stroke, depending on the piston area in relation to the piston rod area.

It is understanding and controlling the hydraulics, the flow of fluid and how much resistance should be built in relative to the piston velocity, making the difference how well a damper operates in the suspension of a race or sports car.

Valving design and controlling the flow of fluid is an important factor.

Fluid passing under pressure through a spring loaded valving system, causes a pressure drop at the dis-charge area which is below the piston in rebound and at the bottom side of the foot valve in compression or in an remote reservoir on the diss-charge side of the valving.

Fluid passing from a high pressure area to a low pressure area causes an increase in temperature, the increase in temperature depends on how high the damping forces are or how fast the damper is moving, (piston velocity) measured in centimetre or inches per second.

This process is in fact the conversion of energy of motion into heat or converting kinetic energy into thermal energy, this is the principal of hydraulic damping and controlling suspension movements.

A damper is a reactive device reacting to displacement and velocity; electronically controlled damping devices are outlawed in most series so we will stick to the most common use of the damper in race, rally and road car use.

There are many myths about damping and how the use of exotic fluids makes a big difference in damping performance; I have not found the need for a magic fluid because it is the valving design which determines the amount of damping at a specific piston velocity.

Viscosity is only a factor when damping forces are generated through the use of orifices, the damping forces are subject to fade due to the viscosity change when the fluid heats up and the fluid gets thinner and more fluid is transported through the orifices.

We start with using a 10 weight fluid with a stable viscosity index.

A stable viscosity index means that temperature increase does not have a noticeable influence on the viscosity of the fluid so the damping forces do not fade with temperature.

The viscosity of fluid passing through a valving system is only a small factor affecting the damping performance when damping forces are generated by the spring loaded blow-off valve, at least not in the damper systems I have designed.

It might be that other manufacturers can make good use of special damper fluids it could very well be but we design dampers with valving systems which do not depend on viscosity changes.

I have always worked with an S A E 10 weight oil with a stable viscosity index which is normal available from one of the major oil companies and never had the need to use more exotic special fluids which are available today because damping forces are generated through means of spring loaded valves in combination with small orifices, the influence of viscosity is neglecteable.

A stable viscocity index is the rate with which the viscosity decreases against the increase in temperature and is measured in stokes or centistokes, the scale shows an logaritmic increase.

The greater volume of fluid is displaced per cm stroke desto easier it is to design a damping characteristic responding to suspension inputs; it gives better control over the damping forces.

A large displacement keeps the hydraulic pressure in the working cylinder lower, a basic understanding of hydraulics and fluid displacement will help understanding how damping forces are generated and can be helpful for any engineer tuning the suspension to make better decisions.

A correct damping characteristic is responding to any movement of the suspension in extension or compression with the correct damping force in relation to the energy of motion generated by the suspension.

This means that the damping characteristic needs to be in line with the suspension characteristic, spring rate, roll rate, velocity ratio, sprung and unsprung mass, which factors have to be taken into consideration.

Damping is basically reducing the energy of motion, allowing dissipation of the kinetic energy over a certain suspension travel.

So that the forces induced by a vehicle in motion, in extension or compression, are not disturbing the cars handling or put severe vertical loads into a driver's body or the structure of a race or sports car.

For sports car drivers and passengers there need to be a good compromise between handling and a reasonable level of comfort.

For the race car driver the comfort level is less important, the handling aspect of the damping input on the suspension is more pronounced.

Information like weight of the car, weight distribution front to rear, spring rates, motion ratio of the suspension, un-sprung weight and kind of tire used are all taken into account for the damper engineer to design the correct damping characteristic.

The damping characteristic needs to provide just enough rebound damping over the velocity range of the graph so that there is not more energy converted and dissipated than necessary for a smooth ride and good handling.

Opinions about the amount of damping applied to the suspension are divided.

There is a tendency among race team engineers to set the bump damping as low as possible, use low gas pressures which leaves the un-sprung mass to move un-dampened, every time the suspension is activated.

This in combination with low spring rates results in a bad handling car.

The use of very little compression damping is not based on any scientific indication but on assumptions, popular believe is that if you use very little compression damping, the car will ride better over bumps and handle better, the opposite is true, the handling will be worse.

The assumption that compression damping should be as low as possible, is illustrated by the following story.

A racing team had been testing a set of JRZ dampers on a F3 car, the handling went better and better by adjusting the damping, increasing the compression damping and decreasing the rebound damping using some higher gas pressures.

After the successful test they concluded that the reason for the improvement in handling was that the arrows with pluss and minus signs on the adjusters were put on in the wrong way because in their mind it could not be that an Formula 3 car ran better with more compression damping.

They did not learn and accept the fact that compression damping (controlling the unsprung mass) is more important for grip and handling than the rebound damping (controlling the sprung mass)

So they send a message to the factory asking to change the signs on the adjusters because they believed that the plusses and minuses signs indicating the increase or decrease in damping were put on the wrongway and should be reversed.

In this book we explain how different damper designs work and how to find the better way to setup a suspension and why compression damping or bump damping is more important for a good handling car than the rebound damping.

Our design philosophy is based on experience in research and development of damping systems and testing in real world conditions, driving cars over bad roads and over stretches of Autobahn in Germany, Highway and Autostrada in Italy at very high speed.

In addition, information gathered at race tracks around the world in all kinds of racing series

Contributes to the understanding of what damping control is needed where and when.

VEHICLE DYNAMICS AND DAMPING

The suspension of a moving vehicle is subject to the input of many forces generated by acceleration, braking, cornering, running on a banked track etc., a correctly tuned damper system can make a major contribution to a perfect working suspension system.

In the following chapters we show the importance of correctly applied damping characteristics, under different conditions, including the differences in damper design.

While being aware that there are other important factors in the complexity of a suspension, playing an important role, we have proven over many decades of damper design and track-side engineering that a well-designed damping system, integrated in a well-tuned suspension system will make the difference offering optimal power-down conditions, traction, lateral grip, better braking performance, better acceleration, faster cornering performance and better tire performance.

We want to explain damping performance in relation to the working of all other suspension components while my personal experience working with race cars, from Formula One, Indy cars and driving and testing the handling of fast sports cars is the guiding line.

The science of suspension and damper design is constantly changing, improving, providing a real challenge for the design engineer, keeping up to date with new developments, meeting the challenge of designing new suspension components that make a contribution to a better and safer auto sport.

03

Damping characteristics

The damping characteristic is the increase in damping force relative to the increase in piston velocity starting from 0 to 5 centimeter, 10 centimetre or 20 centimetre or 0 to 1.96 inch, 3.93 inch or 7.87 inch per second higher piston velocity.

The piston velocity is calculated and expressed as a continues velocity of the piston

Dynamometer testing the damping forces is a great tool to measure the performance of a damper but it should be realized that the dyno sheet does not show the same performance like the damper is working in the suspension of a race car.

With experience it is possible to build the best possible performing damping characteristic for a particular suspension system in relation to the spring rate, antiroll-bar setting and velocity ratio of the suspension, the only restriction being the basic damper design which only allows so much performance.

For example, a suspension in which the damper moves at fifty percent in relation to the wheel travel, the damping characteristic needs to have built-in relative much low velocity damping forces because the damping force needs to increase at a faster rate than the piston velocity does.

On the other hand, a car with a low velocity ratio in which the damper moves one on one with the wheel, the damping characteristic can use less damping in the low velocity range but more damping in the mid to high velocity range so the characteristic needs to be completely different.

VEHICLE DYNAMICS AND DAMPING

A damper working in the suspension of a moving car, does not follow the exact lines of the dyno graph but depends on inputs from the suspension instead of being forced through its travel at a pre-determined velocity by a machine.

Instead of the nice line produced by the dynamometer, the damper in the suspension resists movement as soon as it is activated, slowing down the velocity of the suspension, producing a number of vertical lines alternating above and below the ride height line (zero line) with peaks over bumps and impacts from the road in a frequency equal of the un-sprung mass.

A well-designed damping characteristic counteracts any suspension movement immediately at the very beginning of the stroke, controlling suspension travel by converting kinetic energy into thermal energy, keeping suspension movement within acceptable levels.

The objective of damping the suspension movement in the very low velocity area is to prevent the suspension from reaching those higher piston velocities which are causing a bad handling car.

If the damper controls the suspension movements well in the low piston velocity area, there is no need for increased damping in the high velocity area, so calculating a damping rate expressed in force/meter/second might be an interesting exercise showing a nice even increase in damping force relative to the piston velocity while the suspension assembly in motion might need a complete different damping rate or characteristic.

When a damper is tested in a dyno, the dyno starts its cycle at zero velocity, the damping force is than zero also, when the dyno is building up speed, the damper builds up resistance (damping force).

The damping force depends on the valving design which determines the damping characteristic

In the dyno, the piston velocity does not slow down as a result of damping action like it would in the suspension because the powerful electric motor runs the damper through its cycle keeping the velocity at the pre-set speed so the damper delivers an amount of damping as a response to the velocity and movement of the dyno.

Without knowing the piston velocity in the suspension of a particular car it is impossible to calculate a damping force line from zero increasing to 20 cm per second.

The increasing amount of damping force in relation to the piston velocity is determined by the valving design.

The valving is a combination of orifices, valve shim stack or spring loaded blow-off valve and valve seat area, in case a blow-off valve is used, the diameter of the valve seat and force of the preloaded spring is important for the damping characteristic.

The rebound valve stack is build up from different diameter shims and a specific thickness; I prefer using a double shim-stack with a certain number of shims in the first part against the piston followed by a spacer of a specific diameter and a second stack of shims.

The built up of the double stack can be varied in many different sizes, diameters, thickness to respond to the forces of the suspension input.

Just changing the diameter or thickness of the spacer between the two parts of the shim stack can make a big difference in the damping characteristic and forces.

The characteristic shows the amount of damping force the damper is generating at each point of the piston velocity scale.

In rebound the diameter and total area of piston bores determine how much of the damper fluid can exert hydraulic pressure on the valve. The spring load on the valve determines the moment when the hydraulic pressure overcomes the spring load and the valve opens.

In a damper in motion, pressure builds up in the hydraulic fluid till the pressure multiplied by the valve area are reaching equilibrium with the preload on the valving

This point shows up on the graph as a sharp bend in the damping force line from a vertical to a horizontal line.

Sometimes the graph-line shows a small drop at the very beginning of the graph before continuing, this is the moment that the valve opens, there is a small pressure drop and the graph line drops down a couple of newton's but at the same time the dyno speed is increasing so that the force line drops down only a few newtons.

It shows that the blow-off valve works without leaks and is a positive sign that the valving works correctly.

The moment of valve opening can be varied by changing the combination of valve seat diameter, the spring-load on the valve and how many orifices are open at that moment.

The dyno usually runs a pre-programmed cycle, the machine is programmed to increase the velocity step by step.

The top of the damper is mounted onto a load cell, which in turn is bolted onto the dyno frame.

In the rebound stroke, the load cell records the force, feeds it into the computer processing this information as a rebound or extension force making the damping force visible on the screen as a graph.

Moving in compression, the damping force is recorded by the load cell, fed into the computer and made visible on the computer screen on the same graph, this graph complete with rebound and compression damping forces can be saved and printed for future consultation.

The damping force can be expressed as a force/velocity graph showing the damping force in relation to the piston velocity.

The force displacement graph is a graph on which the damping force is made visible relative to the displacement of the damper piston.

The damping characteristic is the amount of damping increase, relative to the piston velocity, measured in inches or meters per second.

When the damping force increases at a higher rate than the piston velocity, it can be called a progressive characteristic.

When the damping force increase is equal to the increase in piston velocity increase it is called a linear characteristic.

When the damping force increase is less than the increase in piston velocity

It is called a digressive characteristic.

The force displacement graph shows more details of damper performance than the force velocity graph, while most team engineers/people prefer looking at a force velocity graph and most engineers have a preferred line they want the damper graph to follow.

The force displacement graph shows shows clearly the point at which the blow-off valve opens with the temporary pressure drop at the point of opening visible as a small spike at the start of the graph.

The rebound and compression forces are shown as recorded by the dyno, the rebound force on the lower part of the graph and the compression force on the top part of the graph.

In most design dampers, the characteristic changes in the low velocity area when adjusting the damping forces.

That is because the adjustment restricts the flow through the orifices, directly affecting the low velocity damping, increasing the nose angle of the characteristic with every click of adjustment.

The adjustment range of the low speed damping will change the characteristic.

The characteristic can also be altered by changing the combination of orifices, valve stack, preload on valves and fluid displacement, changing a damping characteristic using piston bores of a different diameter is most effective.

The valve stack, a systematic stacking of shims of a certain diameter and thickness is responsible for the damping characteristic at medium to higher piston velocity.

04

Shims

The material the shims are made of is spring steel, Sandvick is a well known brand, it is a stamping product and needs to be without burrs on the edges.

Burrs and knicks cause leaks and preload the shim.\

The spring rate of a shim is very progressive, the first part of deflection is nice and compliant but the spring force shows a progressive increase versus deflection.

For this reason I do not believe in preloading a shim stack unless a specific application calls for it, it seems an easy way of reaching higher damping forces but the shim is preloaded and is more difficult to deflect.

A preloaded shim causes a harsh response from the damper wich is transmitted into the car body resulting in a harsh ride.

Using preloaded shims, a damper with a small piston diameter and less fluid displacement (in general a smaller diameter damper) can produce higher damping forces but does not deliver very nice ride or handling characteristics.

When preloaded shims are used, the first softer part of the spring rate is not available anymore so the shim works in the progressive part of the spring rate scale, keeping the blow-off valve closed longer so that internal pressure builds up to a higher level before the fluid is released.

Shims with a thickness of 0.2 mm or 0.008 inch without preload give the best result in terms of damping response, it is better to use more shims in a stack of the same diameter than less shims and preloaded.

The preload of a shim stack can be so strong that it takes more effort to lift the shims of the valve seat, resulting in a break away force so that the initial part of the damping curve is too steep for a smooth suspension control and ride.

The most used part of damping control by the majority of engineers is the rebound side of the performance curve and results in a harsh ride because the damping force does not allow the spring to extend itself quick enough before it hits the next bump.

The sprung weight will pick up the suspension assembly during the rebound stroke so that the tire is not in firm contact with the road surface.

The initial damping force is responsible for a smoother ride quality of a race car because if well tuned it keeps the tires better in contact with the road surface.

Instead of preloading the shims, thicker shims can be used of 0.25 mm or 0.010 inch to increase the damping forces in rebound or in compression.

This is a better way than preloading, the shim can open freely and discharge a larger amount of fluid when the pressure builds up, so that the initial part of the spring rate is used instead of the much steeper part of the curve.

For damper designs using a small coilspring inside the piston to preload the valve, rebound damping forces can be increased by replacing the coil spring with a heavier one.

There is more than one way to realize a perfect damping characteristic, drivers have different perceptions of handling and driving style, but one thing is clear, all drivers want a perfect handling car tuned to their driving style and the best way to achieve this goal is to understand how to adjust damping in relation to the other suspension components.

In case someone needs to re-valve for the purpose of changing the characteristic and range of adjustment, it is best to contact the manufacturer or the representative present at the race track, or call the manufacturer direct because different manufacturers might have different ways of disassembling and assembling dampers, and with the high gas pressures

VEHICLE DYNAMICS AND DAMPING

present in the damper it might be a dangerous operation for anyone not familiar with the construction of the particular damper.

In a mono tube gas damper with a gas pressure of 25 bar the force holding the top nut or guide in place is for a 40 mm damper 314 kg or 690 pounds.

It is obvious what could happen when the damper is opened by untrained people, first of all the force with which the guide assembly is held in place will not allow anyone to remove the circlip but should some body be able to do that, the damper will explode with a high force spraying damper fluid all over the place and possible injuring the person who is carrying out the work.

Most gas hydraulic mono-tube dampers carry a sticker to not open the damper, and my advice is don't work on a gas damper if you are not trained for it.

So the best thing to do is leave working on the internals of the damper to a specialist with the right training, tools and knowledge how to carry out the work correctly and safely.

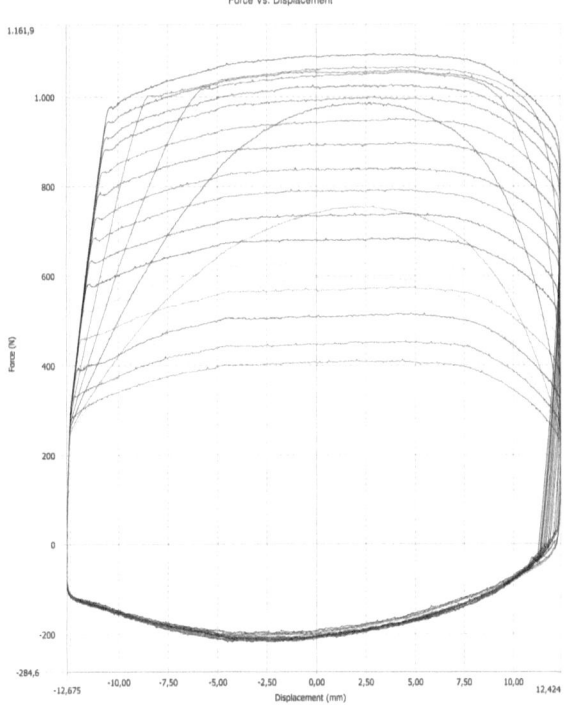

High and low speed compression adjustments
Force vs. Displacement

05

INTERPRETATION OF DAMPER GRAPHS

Both low and high velocity forces are recorded and made visible on the force displacement graph.

The nose angle of the graph is changing with each click of low speed adjustment, the steepest line on the left is adjustment position 5 while the nose angle changes with each click the adjuster is turned down to zero.

Turning the adjuster down to position 4, the high velocity damping is fairly stable remaining at the same value but the nose angle is already changing.

Turning the adjuster down to position 3 there is a significant loss in both high velocity damping as well as low velocity damping as explained in

GRAPH 2.

The damping forces recorded on Graph 2 show that the high velocity damping depends for a great deal on the low velocity adjustment settings, at velocities up to 5 centimetre or 2 inch per second.

With both low and high velocity adjusters at the maximum adjustment position, the nose angle is steep and the damper in the dyno reaches max damping force within the first 5 millimetre of the stroke.

Now leaving **the high speed adjuster** *at* <u>maximum adjustment</u> *and turning* down the **low velocity adjuster** one click, the nose angle changes so that the maximum damping force is reached within the first 10 millimetre of stroke.

The high speed damping is reduced simultaneous with the low speed damping without turning the high speed adjuster

That is because there is a correlation between low and high speed damping forces when the low speed adjuster is turned down, relative more damper fluid is flowing through the orifices, bypassing the blow-off valve.

In graph 2 we have separated the low speed damping function from the high speed function with only the low speed compression damping recorded on the graph.

Leaving the high speed adjuster at the maximum setting, we have turned down the low speed adjuster one click at the time, so that only the adjustments of the orifices are recorded.

It is very important to know that the low velocity adjuster effects the high velocity damping as well.

With all orifices open and moving at very low piston velocity, damper fluid flows through the orifices and only a small portion goes through the blow-off valve.

Starting at the maximum position number 5, the nose angle is very steep and the point at which the blow-off valve opens is visible as a sharp spike, indicating a small pressure drop because the blow-off valve opens and with increasing piston velocity the damping force increases.

The maximum compression damping is reached within the first 7 millimetre movement of the damper stroke in the dyno.

Turning down the adjuster to position 4, the nose angle changes to a shallower angle and the maximum damping force is reached within 11 millimetre of the stroke.

At position 3 while the nose angle is more shallow the maximum damping force is still at the same level but when the adjuster is turned down to position 2, there is a distinct reduction of maximum damping force simultaneous with the nose angle while the piston velocity is still the same.

In Position 2, the high speed damping force is reduced from 1000 newtons (220 pounds) to about 950 newton's (209 pounds force).

Position 1, the high speed damping force is reduced to 720 newtons (158 pounds) and when the adjuster is turned down to 0, the high speed damping force is reduced to 560 newtons (123 pounds)

The reduction of high speed damping at this low piston velocity of 5 centimetre or 2 inch per second is inherent on the hydraulic system, turning down the adjuster from 5 to 0, more fluid is diverted through the orifices, it takes more time to build up hydraulic pressure and to open the blow-off valve.

The fluid volume displaced in compression is relative small and depends on the area of the piston rod and stroke, the fluid flows through the orifices and the blow-off valve system, but all fluid directed through the orifices does not generate damping force at the very low velocity because it is a free passage.

Orifices are open connections bypassing the pressurized part of the cylinder and the low pressure part of the cylinder, with each lower adjustment position more orifices open up and more fluid is deviated through the orifices.

Open orifices mean less damping at very low piston velocity, the damper fluid can flow freely through the orifice in the cylinder from below the piston to the topside of the piston and back, it is the controlled and adjustable fluid bypass which enables the damper designer to manipulate the nose angle of the damper graph.

In the world of motor racing, orifices are a tool to smooth out the sharp edge of the damping curve manifesting itself in a smoother ride but too much open orifice results in greater suspension travel.

Turning the low speed adjuster down, (graph 2) to the 0 position, the maximum low speed damping force is reached at the midway point of the stroke while the high speed damping force has dropped down from 1400 newton's to 400 newton's without touching the high speed adjuster.

This is a phenomenon that race engineers should be aware of when they are tuning the suspension with damper adjustments.

They start with the low velocity adjuster in the zero position, thinking that the car will ride the bumps better, not realizing that the high velocity damping forces are than much lower also. The high speed damping forces are not in the position people think they are.

The lack of low velocity compression damping is because for the un-sprung mass to reach higher velocities than wanted, generating higher levels of kinetic energy, disturbing traction, grip and power down conditions, in the car it feels like the wheels are shaking which in fact it is the whole suspension assembly (un-sprung mass) moving uncontrolled.

The handling of most race cars suffer more from too much low velocity REBOUND damping and low compression damping instead of too much compression damping.

The best way to find the optimal damper settings is to start with the low speed compression damping adjuster in the maximum position which is position 14, leave the low speed adjuster at this position and work with the high speed adjuster first.

Set the high speed adjuster at about three quarters from zero, position 18, the gas pressure should be at 18 bar or 250 lbs.

When the car feels too rigid to the driver, turn down the high speed adjuster first by one click and leave the low speed at the position 14 till the car is responding well to damper settings.

It sounds controversial but the car will respond to bump or compression damping changes in a positive way with better traction, better grip and better lap times.

When the driver finds the car still too rigid, the low speed adjuster can be turned down one click, this will not only change the nose angle but at the same time lower the high speed damping force simultaneous so that one click adjustment has a double effect and will be felt by the driver.

One click adjustment of low velocity damping does not seem to be much of a change but in damping value it is a big step because it effects the initial damping action at the very start of the stroke (nose angle) of the damper while the high velocity damping is reduced at the same time

When the rear tires have no grip sliding in the corner with the power on, one click of adjustment INCREASING THE LOW SPEED DAMPING can make the difference, NOT LESS low speed compression damping like most people tend to believe.

The JRZ damping system is designed to offer optimal ride and handling without being harsh over bumps and curbs.

Graph 3

Graph 3 shows the high velocity compression damping force and adjustment range only, the low speed adjuster remains at the maximum position while the high speed adjuster is turned up one click at the time from the minimum to the maximum position.

There are 14 adjustment positions (clicks); the graph shows the increase in damping force in nice even increments.

The damping forces shown in the graph are recorded at a piston velocity of 5 centimetres or 2 inch per second, which in our experience is the most effective piston velocity controlling the kinetic energy generated by the suspension in motion.

Finding the right mix of high and low velocity damping adjustment is a matter of using a logical approach to suspension tuning, stay away from wild guesses; they usually don't work and can cause total confusion by the driver

Coil spring and damper perform two different functions in the complexity of vehicle dynamics.

A coil spring responds to displacement only, a damper responds to both displacement and velocity.

Below is a damper adjustment graph of a race winning car

The graph clearly shows the compression damping force with a steep increasing nose angle with much low velocity compression damping.

The compression damping is 900 newton's while at the same time the rebound damping is 400 newtons with much less low velocity damping.

06

SPRINGS

A spring is a passive device responding to displacement only, a compressed spring gives back the same force with which it is compressed.

The spring which is compressed during the bump stroke will push the chassis over the ride height level without the controlling rebound force of a damper and keep oscillating

The spring rate is a different story, the spring rate is the increase in force relative to the compression of the spring.

The spring rate is with the weight or mass of the car a component in calculating the natural frequency of the sprung weight

A car whether it is a race car, high performance a sports car or road car is sensitive to the natural frequency of the suspension.

The natural frequency is calculated using the sprung weight and spring rate, there is a formula to calculate the natural frequency.

If the weight stays the same but with lower spring rates, the suspension will use longer travel and the ride feels soft, if the spring rate is too high in relation to the weight, suspension travel will be restricted and the ride feels harsh.

Dynamically, the down force of a car at speed will also have an influence on the natural frequency; the down force acting on the spring will lower the ride height and also lower the natural frequency

Coil springs are available in different designs, the spring most used is the cylindrical spring with a linear spring rate which means that the spring force increase per centimetre or inch is the same over the total length of the spring travel, the coils are evenly spaced.

The wire diameter can be varied, the coil diameter and length can be varied but the size and length depends on the packaging space available in the suspension.

The spring force of the progressive spring increases progressively relative to the compression of the spring, the spring force increases quicker than the deflection.

The coils can be spaced with increasing smaller distances between coils, or some designs use tapered wire to wind progressive springs.

New developments are cylindrical springs with very few windings (coils) like the Eibach company is supplying to race customers.

This design offers a reduction in weight and optimal suspension travel while the spring does not sag or loose its effectiveness under heavy conditions and over its life time

A damper is a passive device, as soon as the damper is activated by suspension movement it reacts to displacement and velocity slowing down the movement of the suspension.

In addition the gas pressure delivers a significant portion of lift (weight carrying capacity) so tuning the damper in line with other suspension components is very important for an optimal functioning suspension and handling of the car.

A spring reacts to displacement only so the suspension velocity does not change the spring force.

Not all coil springs are alike in quality, keeping an equal performance over their life time without sagging or fatigue I have been present at tests where the car would come in after only a few laps and had lowered itself significantly due to spring fatique.

There are differences in quality in materials which are used and the use of computer controlled equipment for the manufacturing process, a science which has been developed over the years responding to the ever higher demands placed on springs.

Once one has walked through a spring factory and observed the manufacturing process, the winding of the springs, hardening, grinding, shot peening and the quality control afterwards has a different appreciation of springs as a finished product.

The mechanism feeding the raw wire into the machine is already a fine piece of engineering.

I have walked through the Eibach factory in Germany several times and every time I have done that I have been impressed with the product design, engineering, diversity, difference in sizes and care with which the products are made and quality controlled.

Designing springs with the correct spring rates within the correct diameters and lengths, choice of materials, performing well over their life time under difficult conditions is a science which has been developed over a long period of time and cannot so easely be copied.

High velocity compression damping adjustment graph each adjustment position is recorded.

High velocity rebound adjustment graph of the four way adjustable dampers.

Each line represents one click of adjustment.

07

Dynamometer

Development of the damper dynamometer started with a simple design consisting of a large heavy steel frame with heavy beams running down the back of the machine to stabilize the frame when running a damper in the dyno.

Moving the frame with a stroke of 75 millimetres, 3 inch and 84 cpm, the machine produced a graph on paper.

The dyno powered by an electric motor, turned a crankshaft and connecting rod, moving a frame in which the bottom end of the damper could be mounted.

The damper industry wide accepted piston velocity standard was 0,33 meter per second/ 1.29 inch per second for low velocity and 0,66 meter per second/ 2,59 inch per second for high velocity.

The top end of the damper was attached by a bracket to a leaf spring.

The leaf spring was calibrated with accurate scales from zero to the maximum force of 500 kg or 1,100 pounds with a linear spring rate, so that damping forces were exact.

The damping forces were recorded on graph paper with millimetre division squares.

A book of graph paper clipped to a holder moved simultaneous with the damper at the same speed and stroke, the insert of a ball point fitted in an aluminium tube was connected to the leaf spring with a linkage writing with ink on graph paper, recorded the displacement of the leaf Spring and thus the damping force.

The graph was not different from what we find on the modern machines of today because we are still measuring the same parameters, the mechanical machine was a useful piece of equipment at the time, it was indestructible and reliable but heavy.

A force velocity graph could be produced running the damper at various strokes and speeds whith the maximum forces of each velocity recorded and plotted on graph paper by hand.

Drawing a line through the points produced a force/velocity graph, which was time consuming since the stroke had to be changed for each velocity but it was accurate since we recorded the damping forces of each velocity at the mid point of the stroke.

By changing the travel of the damper the nose angle of the graph was accurate.

This method gives a more accurate force / velocity graph than the modern dyno's, which are recording the damping forces at different dyno speeds but with the same stroke.

Another dynamometer which was used by the competition and at car manufacturers, was a more archaic type, the machine moved a damper in-between brackets but the lower part of the damper was connected to a cylinder connected with a string.

The string was moving a needle while the roll was turning in phase with the damper, a paper strip was clamped on the cylinder, it was a special kind of paper with two layers, the upper layer could be scratched off, leaving a red line showing the damping forces.

Taking the string off the damper and pulling it, produced a zero line through the damping force lines.

This was still the standard way of dyno testing dampers by manufacturers of high performance sports cars. In the sixties and throughout the seventies.

There was one advantage though.

The only people in the possession of dynamometers were the damper manufacturers and car manufacturers, some machines were located at the damper importer's location or at large distributors but not as widely spread as today.

It meaned that the interpretation of damper performance was done by trained professionals, so that there was less room for speculation and unfounded assumptions because for factory trained R and D engineers it was their daily work.

The R and D engineers of the damper manufacturer were the people looking for new ways, meeting the challenge of the ever progressing science of vehicle dynamics.

Damper engineers need to have a thorough knowledge of the hydraulic processes taking place inside the damper in order to determine what part of the valving needs to be changed for a better responding damping characteristic.

But also, very important, the damper engineer needs to know the effects of damping adjustments on the handling of a car and what part of the damping curve will change the handling, low or high speed.

As part of our R and D work we were servicing the major races like Formula One, sport prototype, long distance racing events like Le Mans.

There is no better and faster education proces for the damper engineer than working closely with drivers and team engineers to achieve the best results in handling and endurance of a race car

There is always the pressure of competition, one moment the car you work with is one of the fastest and it can change in a few moments when your car's position is sliding down on the timing chart.

Performing under pressure is the best indicator how well the engineer understands the complexity of vehicle dynamics and the part the suspension plays.

While working in the USA I was the first one to take a damper dyno to the Speedway in Indianapolis.

It was very helpful in testing the dampers of the teams, showing the crew chiefs/engineers what was really going on when the damper moved in the suspension.

Indy car racing was still USAC and at first we had to convince the organization that we needed a parking in the paddock area for a 40 feet service trailer pulled behind a Ford pickup truck.

It was not easy because USAC was not used to having that kind of service and they made it very difficult to make room for our service truck in the paddock and above all, credentials to access the pits.

The competition, dominating the Indy car series up to this point was not very happy with us either, arguing that drivers or team managers did not need dyno graphs to look at, in as much as the damper rep knews exactly how every different valve would work and drivers would only be confused.

They did not want to or could not explain what kind of role damping played in the car's suspension and handling.

They were wrong, in fact, most people liked the idea that the damper performance could be made visible and within a years' time we had more than 50 % of the field running on the Koni dampers I represented at that time.

In other parts of the world the damper dyno was used for a long time before it was introduced to the Indy car teams at the famous Indianapolis speedway.

Teams like Ford Motor Company came to the Le Mans 24 hours with 16 race cars in addition to the Factory teams Ferrari's, Porsches, Alfa's and the privateers, and all the dampers had to be serviced and dyno tested during the early morning because the team engineers needed to get the cars back together for the next day practice which usually started early in the afternoon and lasted till 12 o clock for night practice to test and adjust the lights.

It was at the Le Mans 24 hour race that the late Carroll Smith and I met for the first time and learned to know each other, Carroll worked with the Shelby Ford team, they were using Koni dampers, the meeting was initiated by Me running into the awning of his race truck with our service

VEHICLE DYNAMICS AND DAMPING

van, Carroll was not the least upset, we became good friends and did a lot of work together for a long time after that.

On one occasion Carroll saved me from being hit with a big hammer by a team mechanic.

It happened before the start of the 24 hour of le Mans when the cars were already in the starting position (the old traditional Le Mans start, in which the drivers were poised on the opposite end of the track, had to run across to the car, jump in, start the engine, fasten belts, close the door and take off).

Carroll came to me and reported that a bump adjuster knob had fallen of one damper and asked if I could put it back on in the right position without taking the dampers of . . .

The aluminium knob was secured with a set screw, but the aluminium material expanded more when the damper heated up than the steel shaft and worked itself loose.

So we had the car jacked up and I crawled under the car to put the knob on in the right position.

While I was working under the car Carroll got distracted and walked over to the other car.

A team mechanic who was not aware of what I was doing got the impression that I was sabotaging the car or something to that effect.

He was shouting something to me but being under the car, I could not hear Him very well and did not understand what was going on till Carroll noticed the commotion and came running back stopping the guy from hitting me.

Later on the mechanic and I met in different places at race tracks, I could not help but reminding the man of his action.

In fact when Carroll Smith worked with Moffatt in Australia and I was Chief Engineer at Koni in Hackensack New jersey, Carroll did not hesitate to call me in the middle of the night at home discussing handling problems with his cars, I never did mind getting up and discuss His problems, it was always nice to talk to Carroll again and I liked His relaxed way of discussing solutions.

Jan Zuijdijk

Later When Carroll was working on his first book "Tune to win", He asked for detailed information about dampers and damping for race applications

In le Mans there were many factory teams like Ferrari, Porsche, Alfa plus the private teams etc., the factory teams bringing minimal four or more cars to the race, it is evident that a lot of damper work was done in our service truck at the track from dyno testing to re-valving and overhaul, some private customers waited to service their dampers till one of the big events because it was convenient and the service was free of charge.

The dyno was upgraded with a processor and scope, the graph was made visible on a small screen, not much of a difference in itself but it looked more professional since we were entering the electronic age, people liked to look at screens.

A new development was the dyno with electronic processing of the damping forces by a computer, this made work a lot easier, the machine can be programmed to run through a set of different increasing velocities, producing a force velocity graph instantly.

There were hydraulic driven dynamometers available but they were huge, heavy and not portable but capable of heavy loads mainly in use to test railway dampers when I worked at Koni.

Those machines were operated by hydraulic actuators requiring large capacity hydraulic pumps and high pressure reservoirs keeping a supply of hydraulic fluid under high pressure available.

The machine needed a special foundation to dampen the vibration and shock waves of the machine in operation.

It was equipped with a 16 channel tape player which could record the damper movements, taped in the real world of running on a track into the machine and it would reproduce all suspension movements exactly as they occurred with a damper mounted in the machine while we could watch, observe and measure anything that was going on with and in the damper.

As a matter of fact, at one point Amtrak had problems with railway dampers on a train in the USA, the axle dampers kept disintegrating in a very short time, on my request the suspension of the train was instrumented with transducers, the train made its daily run recording the data of the axle dampers on a 16 canal tape.

VEHICLE DYNAMICS AND DAMPING

After the testrun I hand carried the tape back to the factory in Holland,

The tape was played on the heavy MTS dyno and the machine reproduced all the suspension movements including the usual vibrations occurring in the suspension of a train when braking etc.

We installed the same type damper in the machine, after a certain time the damper showed the same failure as in practice under the train although the problem was not the damper but the different track conditions, while the railways in Europe used welded rail, In the USA they were still using the rails bolted together with a small gap for extension and retraction due to temperature conditions.

With this information it was relative easy to make design changes to solve the problem.

This MTS dyno was installed and operational long before the portable dynamometers were available for the race teams.

The electronic dynamometer for race dampers became readily available when Kurt Roerig designed a small portable machine with electronic processing which could be mounted in a transporter and taken to the race track or used in the factory.

It was still the cranck and connecting rod type but with electronic processing of the data.

Now people had a tool to evaluate the dampers they were using and make intelligent decisions and adjustments so it seemed.

I think that people were more confused than ever, because if one is not used to make the right interpretation of damping graphs it is very easy to read more into it that there really is or make the wrong interpretation.

To make the right interpretation, one has to understand the hydraulic processes taking place inside the damper, at JRZ Suspension it is our every day business to design and understand the hydraulic processes.

There are as many different opinions as there are users as to what a damping curve should look like for a particular car and what will work and what not.

Many times, people, when switching to another brand of damper ask to copy the damping graph of the brand they were using before, they think that the damper will perform the same.

They have to understand that this is too simplistic because copying the graph does not mean that the damper has the same effect on car handling, the damping system might be completely different and therefore the effect on the handling of the car will be different.

People often overlook the fact that different manufacturers use different test velocities so the graph might look the same but the damping characteristic is not the same and the effects on handling are not the same.

Damping forces should always be read in relation to the piston velocity, look at the characteristic and check the scale of the graph.

If piston velocity, damping value and scale are not recorded on the graph, the graph is useless.

The best way to find a better setup is to watch the stopwatch, observe the car and listen to the driver.

The behaviour of the car and the driver's report will tell more what the car needs than a predetermined line of a damper graph.

If the driver reports under-steer or over-steer conditions and the stopwatch indicate slower lap times than I assume that there is room for improvement in de handling of the car.

I have been in situations overhearing the communication between driver and team engineer in which the team engineer was so forceful putting forward his own opinion that in the end the driver was totally confused and the wrong changes were made.

Beware of a situation in which a nephew or uncle of the driver is present at the test and comments on the handling at what He see the car do wrong, I had that happen several times, they have no idea what your plan is but confuse the issue by intervening in the process.

I always have had confidence in the race drivers with which I worked the test with and the equipment they have to work with.

Assuming that the car setup is the limiting factor, I take the driver reports as the base to make the decision about set up changes.

Emerson Fitipaldi, Indy cars and Ron Fellows, Camaro chassis development program were two outstanding drivers who gave detailed information on car handling and what they wanted to be improved.

The shape of a damping graph depends for a great deal on the damping system which the manufacturer is using, so the graphs might look alike but they have a different effect on ride and handling.

A damper is a reactive device, when not moving it does not produce any force; the only force noticeable is the gas pressure if the damping system is pressurized, extending the piston rod.

Gas pressure exerts a force on the cross section of the piston rod, extending the rod with a force equal to the pressure multiplied by the area of the rod.

It cannot be read from dyno graphs how good or bad a damping characteristic is for a particular car without knowing the settings of all other components like spring rates, antiroll bar kind of tire etc. but when a damper is not functioning correctly it will show up immediately on the graph.

The dyno graph is very useful for the engineer comparing the performance from one damper to another and check the damper settings but don't read more into it than there is.

The dyno graphs show the damping forces in relation to the Piston rod velocity with the adjustment settings and valving at which the damper is set at of the moment of running it in the dyno.

The force velocity graph shows the damping characteristic which is the increase in damping force in compression and rebound relative to the increase in piston velocity starting from zero till the max velocity showed on the force /velocity graph.

It should be realized however that a damper operating in the suspension of a car does not follow the exact line showed on the graph of a dyno; suspension travel and velocity in the real world are totally different from the pre-set program of the dynamometer and the graph it produces.

A suspension in motion does not follow the nice 5 or 10 cm (2 to 4 inch) stroke at a certain preset velocity, it moves intermittent with the natural frequency of the un-sprung mass and sprung mass with a stroke depending on the kind of road surface on which the car is running.

The suspension of a rally car running on dirt is subject to much longer travels and a higher piston velocities than the suspension of a race car on a road course, damping characteristics need to be adapted to the application.

Transducers mounted parallel to the damper give a good indication about damper movements, stroke, velocity, and frequency of the damper at certain tracks.

Whereas a damper working in a suspension slows down suspension movement by applying a certain dose of damping responding to velocity and severity of a bump, a dyno forcefully moves the damper through the stroke and velocity programmed by the operator, the dyno is not slowed down by the damping force so the damper performance is recorded at the velocity and displacement the dyno is programmed at.

With each stroke the dyno moves the damper from zero piston velocity to the programmed speed at midpoint of the stroke, past that midpoint the velocity slows down to zero again and the process is repeated till the dyno is switched off or run at a different speed.

Let's say that the dyno is programmed to run at a speed of 20 cm or 8 inch per second.

When the dyno is started to run, it starts at zero velocity pulling an pushing the damper through the programmed cycle and speed, the maximum damping force is reached at the midpoint of the stroke in both the rebound and compression cycle.

The damper responds with a certain force in rebound and compression which is recorded by the dyno's computer and displayed as a force velocity graph or force displacement graph, showing not only the damping force but also the rate with which damping force increases in relation to the piston velocity.

Also the part of the graph in which the dyno slows down to zero velocity is important to watch because it will tell the user much about the low velocity damping.

The piston velocity of the dyno can be calculated using the formula

Pi X stroke in cm X RPM divided by 60 = meter/second velocity. for example 3.14 X 0.05 m X 80 rpm: 60 = 0.209 meter/second

Or in inches per second 0.209 m/second divided by 2,54 = 8,22 inch / second

One inch is 25.4 mm, 2,54 cm or 0.0254 m.

That is when the damper runs in the dyno at a pre-set piston velocity

A damper operating in the suspension does exact the opposite as what the dyno curves show.

A damper mounted in the car reacts to suspension displacement (travel) and velocity of the suspension while a damper in a dyno is forced through the cycles.

When the suspension is activated by running over a bump or dip in the road surface, the low velocity damping reacts immediately by resisting movement at the very beginning of the stroke in rebound or bump, slowing down that motion keeping the suspension travel as minimal as possible.

Depending on the force input and damping characteristic of the damper the suspension will slow down in a controlled way keeping the tire contact patch in contact with the road surface.

CRITICAL DAMPING.

One hundred percent rebound and compression damping, meaning the low velocity damping force is so forcefull that any movement of the suspension is blocked by the damper.

This is called critical damping, in this situation the suspension is rock-hard, the car will be jacket down, lowering dynamic ride height and the tires will lift of the road surface at every bump or dip destroying grip and traction.

It is clear that this is a situation which nobody wants.

A correct damping characteristic allows the suspension to move at a controlled velocity and travel, that is why damping forces should be independently adjustable in low and high piston velocity.

If the damping characteristic shows a steep nose angle, indicating the damping force increases quickly at low velocity, the suspension movement will be slowed down immediately with a greater force than when a characteristic is used with less low velocity damping.

This is very important because the handling of a race car depends for a great deal on the damping characteristic in combination with spring rates, roll rates, weight and wing settings.

Transducers measuring suspension travel, are mostly mounted parallel on the damper and show the suspension travel and frequency for each wheel.

It is interesting analizing data taken of a car, transducers give information about suspension travel, travel caused by roll, acceleration and de-celleration, braking and running over a curb.

The data should be read in relation to the part of the track the car is running on which is possible by moving the cursor over a map of the track so that the suspension can be seen working at a particular part of the track.

A car running over a bump under braking will show greater suspension movement than when running on a straight line which is visible in the form of a spike in the data.

When looking at data, I take into several factors in consideration which are directly related to suspension performance.

Running over a curb will be recorded as a sharp spike, braking for a corner will reveal grip levels, the rear suspension will open up and the rear tires get lighter while the front suspension will compress (dive) at the same time and the front tires loaded heavier so there is a temporary imbalance front to rear.

This situation can lead to bad handling characteristics.

Increasing the gas pressure of the front suspension can improve this situation.

VEHICLE DYNAMICS AND DAMPING

I have been in a situation in which the team had modified the rebound valving of the rear dampers they were using, the rebound of the rear dampers was very strong, they had closed off some bores through the piston believing that it was the ultimate improvement.

The driver was slow; He reported the car being very twitchy under braking and lots of over-steer when accelerating off the corner.

The team did not tell me about the modification but the driver was not happy about the handling at all.

Obviously there was something wrong, opening the dampers after practice revealed the problem they had created for themselves.

Bypass holes closed off and way to much spring load on the shim stack.

Restoring the valving to the original setting was the solution to this problem, after we restored the dampers to specifications, the car was very good under braking and the over-steer was gone, in fact the handling had so much improved that the driver by the start of the race shot from 27th position to the 12th in the first lap and was working his way to the front of the race when a car crashed in front of our car, crossed the track and crashed into the car which was the end of the race but we had proven that the correct damper characteristic can make a difference of night and day.

Acceleration shows the grip and traction level of the driven tires front or rear wheels.

Compression and rebound movements reveal cornering (roll) when one side compresses and the other side opens up.

Working with a Formula One team in Monza Italy, I noticed that the data showed the rear end of the car was sagging below ride height at speed, My conclusion was that the spring rates were to low and the springs could not support the rear-end of the car

Running in this position, the dynamic settings do not work either because they are set with the car at the proper ride height.

As a result the car lost much of its aerodynamic function at speed and was difficult to drive causing the driver to spin in several places at the track.

When I pointed this out, the technical staff did not agree except Richard Divilla, a Brazilian race engineer who was working as a consultant with the team, Richard agreed with my conclusion that the spring rates were to low and the rear suspension was compressed, lowering dynamic ride height.

The rear spring rate was increased the car started to run much better and the driver was able to compete and qualify, which they could not do before.

It shows that the best data is not worth anything if the right interpretation is not made.

The dyno is a good instrument to measure damping forces in relation to the piston velocity of the damper but finding the setup for the best performing car is still a matter of considering the input of all suspension components making the right decisions and a correct mix of adjustments.

Once a good race setup is found and all suspension components like spring rate, antiroll bar settings, dampers are adjusted to peak performance, the dampers should be tested on the dyno and the damping curves recorded as a reference point but always in combination with the values of all other suspension components.

When components like spring rate or anti roll bar settings are drastically changed, the same damping curve might not work so well.

What the dyno does not do is giving a prediction of what damping forces or characteristic would be best for the car and track.

I have worked with engineers who declared the damping characteristic sacred and relied completely on the way the damper followed the line of the damping curve, they have been lead to believe that calculating the damping rate in newton's per millimetre per second is a sure prediction of what the car/suspension needs for good handling.

I have never bothered making calculations to find the damping rate per meter /second, it is all theoretical, very time consuming and the reality of the car running on the race track throws every calculation off the minute the car leaves the pits

One can make the calculations in the office but everything changes when the car is on the track and you have 45 minutes to qualify, there is no time to make lengthy calculations but you have to think on your feet, analyse the situation in your mind and make decisions that improves the handling instantly.

The problem is that the track surface is different at every track, every inch of the way while drivers may have different driving styles.

Keeping a dyno curve at hand is very usefull because when a good setup is found and the damping recorded on a force velocity graph, it is good to have a reference.

The simple fact of increasing the low speed compression damping results in different, shorter suspension travel changing the behaviour of a car without sacrificing any grip.

The dyno can be programmed to show different damping graphs, when studying a graph always take into consideration the stroke, rpm and the scale with which the forces are recorded.

Most people prefer looking at a force velocity graph believing that this graph gives better information but for damping performance the force displacement graph shows more detail.

The force line starts at zero velocity and increases gradually with the increasing velocity to the highest value till max velocity is reached in both rebound and bump; it shows the cycles from start of the up cycle to the midpoint and down cycle after the midpoint of the stroke.

Closing orifices on the rebound side results in more direct low velocity damping which is not very good for traction

More shims or more preload on blow-off valve result in higher damping forces at the mid to high velocity range.

A damper is most effective when the damping forces respond well with a steep increase in force from the 0 centimetre to 5 centimetres per second velocity or from 0 to 2 inch per second.

Different manufacturers use different parameters, testing the performance of the dampers, but experience learned that when dampened well in the 0 to 5 centimetres or 2 inch per second velocity range, most cars respond well to the damper adjustments.

Damping forces at higher velocities are less important.

That reading a damper graph by people who are not completely sure what they are looking at illustrates the following exchange between a team and a damper/suspension specialist.

The team tested successfully with a new set of dampers with external reservoir, a so called double adjustable damper with external reservoir.

Before the test, they were given the damper adjustment settings and recommended gas pressure to start with.

They did very well in the test, they were the fastest car all weekend and won every race in their category, the problem arose when they went home and examined the data of the car.

Because they were using a different system, the data looked different than what they were used to and jumped to the conclusion that it could not be any good.

First they claimed that the suspension never moved below the ride height level and they went on to say that they could not compress the suspension when standing on the back of the car, with two people and jumping up and down.

Now this does not mean much because you need two people to move the car because one has to compress two springs in addition to the gas force, of course they called and explained what they had done, I told them that they were wrong in their assumptions and should look at it again.

A week later they called again to apologize and said that they had read the data wrong, instead of the car not moving below the ride height line, the car moved ONLY below the ride height line and not over it, so it goes to show that wrong conclusions are easy made.

Now instead of the suspension being too stiff, it turned out that the suspension was on the soft side and that they could have done much better if they had followed our advice with the damper and gas pressure adjustments.

It is my experience that a race car improves in traction and handling with sufficient gas pressure to help dampen the un-sprung weight.

VEHICLE DYNAMICS AND DAMPING

The displacement/force graph gives a lot more information about the operation and performance of the damper in both rebound and compression than the force velocity graph.

The nose angle of both rebound damping and compression forces are better to read because the graph shows a much better line from the zero point to the point of blow-off than the force velocity graph.

The different lines of adjustment are much better defined and visible.

Any free travel (when air has entered the system) shows up immediately when the graph line follows the zero line for a short time and turns sharply up.

A small amount of Air in a gas pressurized damper is not so dramatic; the air bubble will be compressed and diffused into the fluid as soon as the system is pressurized.

Any hydraulic fluid contains a small percentage of gas which is diffused in the fluid but the percentage is so low that we will never notice.

It is only visible when oil and gas are pressurized and the pressure is released; any air present in the fluid comes out as foam or small bubbles when the damper is opened.

It is not so that the performance of gas pressurized damper is affected, as long there is pressure in the system the damper will perform well because the gas or air is compressed into the fluid.

This is only true when higher gas pressures are used, for lower pressures than 6 bar or 100 lbs this does not work.

Air or gas mixed with damper fluid in the system can affect the operation of the damper.

Gas pressure always needs to be separated from the fluid.

Another cause for the dyno to follow the zero line for a short moment can mean that the brackets with which the damper is mounted in the dyno have some play or if the damper does not fit perfect in the brackets.

The force displacement graph shows also the correlation between low velocity damping (nose angle) and high velocity damping.

When both the high and low velocity adjuster of an adjustable damper are turned into the maximum position, the graph shows the nice equal increase in damping force with each click of adjustment.

Turning the low velocity adjuster from maximum to minimum when running in the dyno the nose angle changes from very steep to a more shallow position but it effects the high velocity damping at the same time.

Each time the low velocity adjuster is turned one click lower, the high velocity damping is reduced simultaneously with the low velocity damping.

When someone at the race track turns the low velocity adjuster down one click it automatically reduces the high velocity damping force and the driver feels it.

It should be realized when tuning the suspension at the track that turning down the low velocity damping force adjuster to the zero position, the high velocity damping force is reduced to almost zero also at 2 inch or 5 cm per second.

It is therefore important to know the average piston velocity occurring at a particular track and keep in mind that low adjustment positions from 0 to 2 of the compression damper settings do not improve the handling but make it worse.

Once that is the case, adjusting other suspension components like spring rates, anti-roll bar settings etc. don't have any effect on the handling.

Most on car data acquisition systems show the suspension movement and an average damper velocity of the car running for each part of the track or one can take a time laps of one second and count the number of times the damper has moved.

Experience has learned that damping is most effective at the 0 to 5 cm/meter or 2 inch/meter piston velocity range.

One click of low velocity damping can change the handling of a car completely because the damping is very effective at that low velocity.

08

Twin-tube damper designs

The twin tube damper design is still available from many damper manufacturers, it is a well proven design that works well for road cars and racing series like show room stock and race series with restrictions on suspension modifications.

Cylinder and reservoir tube (damper body) are placed concentric, the inner tube is the working cylinder in which the piston is riding, piston rod and piston are responsible for fluid displacement, moving fluid in and out of the cylinder into the outer tube which is the reservoir space for fluid and air.

The air in this design is not pressurized but in later designs replaced by gas under pressure.

The piston diameter in a twin tube damper is smaller than in a mono-tube because the outside damper diameter has to be kept within certain limits in order to fit inside a coil spring. Whereas a mono-tube can use a 40 mm, or 1,57 inch piston, in a twin-tube damper with the same outside diameter the piston diameter cannot be larger than 25 to 30 mm, 1,01 to 1.18 inch.

The piston valving is responsible for the damping forces when the damper is moving in extension, building up pressure in the upper part of the cylinder, releasing damper fluid through the rebound valving system into the lower part of the cylinder while drawing fluid back into the cylinder through the one way valve of the foot valve at the same time.

On top of the piston is a one way valve, closing the flow from the upper part of the cylinder to the lower part during the rebound stroke so that all fluid passes through the orifices and valving system of the damper.

During the compression stroke, fluid passes from the lower part of the cylinder to the upper part in order to fill the upper part of the cylinder.

The piston rod volume responsible for fluid displacement in compression displaces the fluid through the valving system of the footvalve, generating the compression damping.

The foot valve usually incorporates one or more orifices and a blow-off valving system.

On top of the foot valve is a one way valve, letting fluid through only during the rebound stroke and is closed during the compression stroke, the valve is spring-loaded, preventing fluid from bypassing the valving system.

A leaking non return valve can cause a loss of low velocity damping, which is visible on the graph as a shallow nose angle, the leak might disappear when the dyno is run at a higher speed but it will return at the very low velocity.

When this occurs the damper has to be repaired or the foot valve replaced.

In a coil over design the outside of the damperbody can be threaded and functions as spring carrier, using threaded spring platforms to make the ride height of a car adjustable by turning the spring platform, moving it up or down as necessary.

This feature comes in very handy when adjusting corner weights.

Twin tube dampers function well, the drawback is that not every brand of damper is external adjustable and has to be taken of the car to revalve either rebound or bump valving.

This process is time consuming, elapsed time between removing the damper from the car, changing the valving and re-installing the dampers on the car is too long for a driver to be able to compare the performance and handling of the car before and after the valving change.

The reservoir tube, is connected to the lower part of the suspension and the piston rod /piston assembly is connected to the chassis or car body.

For racing applications, the rubber mounting bushings are replaced with bushings of different material.

Twin tube dampers without an external fluid reservoir cannot be mounted upside down; the foot valve at the bottom of the cylinder is drawing fluid from the reservoir and has to be at the lowest part of the damper so they have to be mounted with the piston rod on top.

In most damper designs, the body can be threaded to fit adjustable spring platforms, other designs use grooves machined in the reservoir tube which are evenly spaced apart to adjust ride height in even increments of 10 millimetre, In this case the spring has to be removed

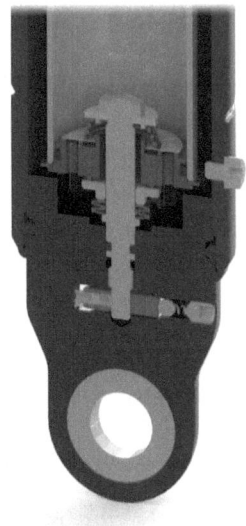

from the damper body to change the ride height.

Twin-tube dampers are available in different size piston diameters.

The piston size depends on the damping forces which the damper is expected to generate.

The fluid displacement in rebound is 3,9 centimetre square,1,53 square inch this is a significant smaller working area than the mono-tube damper with the same outside diameter.

The piston rod diameter in the twin tube design is 12 or 14 millimetre or 0.45 inch 0.55 inch

The fluid displacement of this size piston rod is 1,13 cubic centimetre, respectively, 1,54 cubic centimetre or 0, 18-0, 24 cubic inch.

In most twin tube dampers, the top of the cylinder is closed off by the guide, the top of the cylinder is sealed by metal to metal line contact pressed against the guide body and at the bottom of the cylinder against the foot valve.

The cylinder needs to be pressed tight against the guide and foot valve, therefore the top of the damper body is threaded on the inside using a nut to tighten the cylinder against the foot valve at the bottom.

A radius is machined on either end of the cylinder so that there is only a line making contact with the guide and foot valve, sealing it tight.

Anybody taking a twin tube damper apart should make sure that the radius is clean and intact without dents to prevent internal fluid leaks affecting the damping performance, when dents are visible the cylinder tube should be re-machined making a nice new radius on the ends, taking of a little material on the ends is not a problem, be careful clamping the cylinder in the lathe, the wall thickness is only 1,5 millimetre, it is easy to dent the cylinder rendering it useless.

This way of sealing is a proven concept, providing a tight connection between cylinder, guide and foot valve.

In the present designs, o ring seals are used to seal the cylinder.

The rebound force is generated by the piston area, the fluid volume displaced by the working area of the piston builds up hydraulic pressure between the top end of the piston and the guide, releasing fluid through the rebound valving into the lower part of the working cylinder at a pre-set damping value.

At very low velocity fluid is displaced from the upper part of the working cylinder to the lower part through one or more orifices which are fed through a passage inside the piston rod.

With increasing velocity, pressure is building up and the blow-off valve releases fluid into the lower part of the cylinder, the piston rod volume has to be drawn back into the cylinder through the one way valve in the foot valve.

In compression, damping forces are generated by the displacement of fluid by the piston rod volume, when the damper starts moving at very low velocity, fluid flows through one or more orifices, this is the initial part of the damper curve.

The flow characteristic of an orifice is very progressive relative to velocity, at very low velocity fluid can flow unrestricted through orifices but with increasing velocity the fluid volume becomes too large for the orifices to handle, the damping force increases rapidly and the blow-off valve of the foot valve takes over releasing fluid into the reservoir

In the compression stroke pressure builds up in the entire cylinder, above as well as below the piston, at the point that the blow-off valve opens, fluid flows from a high pressure area to a low pressure area producing heat.

This is the point where kinetic energy is converted into thermal energy or heat in any damping system.

The heavier the damping force the more heat is generated which has to be dissipated into the airstream.

The heat is transported into the reservoir tube, which is in contact with the outside air cooled down by air striking the damper body surface.

The next step up in damper size is the so called heavy duty damper, the piston diameter in this design is about 33 millimetre 1,3 inch and the piston rod 16 millimetre 0,63 inch

The fluid displacement in rebound of the 33 millimetre piston is 8.54 cm square minus the piston rod area which is 2 centimeter square leaves a working area of 6.54 centimeter square.

The piston rod displacement in this design is 2 cubic centimetre 0,32cubic inch for each cm stroke which is twice as much as the 12 mm piston rod diameter displacing 1 cubic centimetre fluid per centimetre stroke in compression.

More fluid displacement per centimetre stroke means better control over the flow through the system and therefore better damping performance at low piston velocity.

The operation of the heavier damper is the same as described before, only the damping forces are higher but the internal pressure remains almost the same because the damping force is generated by a larger piston and piston rod area.

The rebound damping force can be changed by modifying the preload on the blow-off valve, changing the flow through the piston or use a heavier shim stack on the blow-off valve.

To modify the valving the damper needs to be opened with a special tool, in most designs there are blind holes in the top nut in which a tool with two pins can be inserted to take the top nut off, sometimes the top-nut has a hexagon profile which can be used to open the damper with a big wrench but before you do anything, LET TE GAS PRESSURE OUT.

IF A GAS DAMPER IS NOT EQUIPPED WITH A SCHRADER VALVE TO RELEASE THE GAS PRESSURE, DO NOT OPEN THE DAMPER BUT LEAVE IT TO THE MANUFACTURERS REPRESENTATIVE OR A QUALIFIED REPAIR SHOP.

Some twin tube dampers are pressurized with hydrogen gas, replacing the atmospheric pressure with 6 bar of gas pressure, this system is called soft gas or low pressure damper.

Be careful not to open a damper with gas pressure inside, check for gas pressure first and release it before opening the damper and make sure that the damper can be pressurized through a Schrader valve again after completing the modification. When the damper is not equipped with a Schrader valve, it can be that a small rubber plate is used through which a needle can be inserted to fill gas, the same special tool is needed to deflate the gas pressure.

When the nut comes off, the cylinder filled with oil can be lifted out of the damper body complete with footvalve which is clamped inside the cylinder.

Be careful not to spill the fluid which is still in the cylinder because it is needed to re-assemble the damper again after the revalving has been completed, if there is no other damper fluid available, but it is prudent to run the fluid through a sieve cleaning out any particles.

With the piston-rod assembly out of the damper body, examine the piston valving located underneath the piston and see whether it can be dis-assembled.

Sometimes the nut on the piston rod is staked to prevent loosening and cannot be taken off.

If that is not the case, see whether the preload on the valve can be increased (an extra shim might help) or another blow-off valve can be used.

Some manufacturers recommend shim stacks for different track situations.

Changing the damping force of the foot valve might not be so easy, most dampers have fixed valves of which only the preload can be changed.

Reassembling a twin tube damper is not difficult; make sure the foot valve is installed in the cylinder, in most cases the foot valve clamps lightly into the cylinder so that it stays in during assembly.

The cylinder with foot valve attached is placed inside the damper body in the centre and completely filled with fluid while the reservoir (outer tube or damper body is filled half way or a little less than half way).

The space around the cylinder in the top half of the outer tube is needed as air or gas reservoir to accept the fluid displaced by the piston rod volume.

To avoid all this labour send the damper to a repair shop or select an adjustable damper with an extended adjustment range if the rules permit to do so.

09

Development Of The Twin-Tube Hydraulic Damper With External Adjustment Knobs

Development of the external adjustable damper was one of the most important developments for all racing series active in the racing community and still is

Racing dampers were not external adjustable at best an upgraded road car damper was used.

In the early sixties, after testing at the Nurburgring with the Porsche Factory racing team, we returned home for the weekend leaving our service van in Germany.

Henk Richten and I took a train back home leaving the service van parked in a safe place in the city of Bonn, Germany.

During the train ride back home we had plenty of time to discuss the test results and we came up with the idea of making the twin-tube koni dampers externally adjustable, so that we did not have to take the dampers of the car for a valving change and save a lot of time running the car instead of waiting in the pits for the dampers to be revalved.

We decided that the best way would be to make the rebound valving adjustable through the hollow piston rod so that we could turn the adjuster nut underneath the piston with the dampers still mounted in the car.

VEHICLE DYNAMICS AND DAMPING

It has been said that making the piston rod hollow was de most important development of the race damper and we should have patented the hollow piston rod, only we did not know that then.

A problem was that we did not have hollow piston rods; up to this point in time all piston rod material used in the industry was of solid material.

Having an idea is fine but making it work is more complicated and a complete different story.

I started working with the only engine lathe we had in the department trying to drill a hole in a piston rod lengthwise.

Not to make it more difficult than necessary I started out with a short rod.

After several failures due to broken drill bits coming out of the side or other problems, I had one which I could use.

We did not have drill bits long enough to drill through the entire length so I had to make them myself.

The next job was to design and make an adjuster rod fitting in this hollow rod and seal it at the top and bottom preventing fluid leaks.

The connection with the adjuster nut was not too difficult, this part done in a fairly short time; the difficult part was how to operate the adjuster shaft while having a mounting eye fitted on top of the rod.

I made grooves in the top of the adjuster shaft and using parts of a radiator hose clamp, having a worm drive installed so we could turn the adjuster shaft with a screwdriver.

When this part was done we turned our attention to the compression side of the damper.

The bump adjuster was more difficult to design because the compression valving in the foot valve was always set at a flow bench in the factory, making it adjustable we needed to start from scratch.

My college, went to the local watch maker in Our Beyerland and brought back small sprocket wheels from an alarm clock, I Souderd the sprocket wheel onto the little bolt holding the shims in place.

A hole was drilled in the side of the reservoir tube so we could install the little drive shaft and sealed with an o ring seal.

The first time I put this prototype external adjustable damper on the dyno it was really amazing how well it worked.

All of a sudden we could adjust bump and rebound forces independently from each other with the turn of a knob without taking the damper of the dyno.

In that very moment, Mr A de Koning walked into the room and asked what I was doing all by myself in the research department during vacation time.

Mr Arie de koning was senior director and owner of the company together with His brother Kor de Koning, founders of the company, He knew me well since i traveled frequently with both Directors but being there al by myself in vacation time was not normal procedure.

I showed what the damper could do.

He said come lets go to my office and explain to me what it is and how you made it work.

Mr de koning being a hands-on Engineer had a fine sense for new developments and saw the importance of the external damper at once because this design would change the way racing dampers were build and used forever.

We discussed the design of the damper and how to improve the basic design.

When everybody returned from vacation it was my turn to take a vacation because I had to fit it in with the racing schedule.

After I went on vacation, the R&D department made a different version of the compression adjuster because my version with alarm clock parts was not very practical or adaptable for production, it was only designed to see whether it would work, the engineers who took over the project came up with another solution consisting of pulling a tapered wedge under the shims of the foot valve, preloading the shim, increasing the compression forces.

VEHICLE DYNAMICS AND DAMPING

This mechanism adjusted the compression damping but there was so much play in the system that it was highly inaccurate and not useable.

We decided to redesign the compression adjuster.

Drilling a hole in the cylinder close to the bottom I soldered the blow-off valve seat direct onto the cylinder, separating the blow-off portion of the foot valve from the part containing the one way valve, keeping the one way valve assembly in place at the bottom of the damper.

We designed a compact adjuster unit with a click function for accuracy, and the external adjustable twin tube damper was ready for production

A new production line for race applications was put in place.

So it goes with many new developments, it starts with not being satisfied with what is available at the moment, finding a better way to do the job, from this first design, adjusting the rebound force through the hollow piston rod several more applications were derived, to mention one, the load dependent damper for buses with air suspension, using the available air pressure in the air bellows to adjust the rebound damping forces.

The air pressure in the air bellows varies because when people walk in or out of the bus the load varies also.

Using this pressure variation, the damping adapted to the load condition of the bus, the load dependent bus damper had a diaphragm on top of the piston rod which was connected with a pressure hose to the air suspension of the bus, the air pressure variation in the air bellows adjusted the rebound damping, increasing or decreasing the preload on the blow-off valve and closing off small orifices at the same time.

When we had a more industrial developed design of the external adjustable damper for race car application, we tested the damper for the first time with Ferrari at the North-sleiffe of the Nurburgring, the famous original race track in the Eiffel Mountains.

The cars arrived direct from the 1000 kilometre race in Franchorchamps, they had won the race, the test at the Nurburgring was in preparation for the 1000 km race of the Nurburgring two weeks later.

The cars were driven by Ludovico Scarfiotti, Mike Parks and Lorenzo Bandini, world famous drivers, with lots of experience in Formula One as well as sport prototype cars.

The test was a great success because the time spend in the pits making adjustments was at a minimum, and the car could leave the pits after an adjustment in a couple of minutes instead of hours.

Mr Franco Lini, a journalist writing for the Italian auto sport magazine "Quattro Ruote" was present at the test shooting pictures and writing a story about the new development in his report, which I still have in my collection.

After this successful introduction, the racing community quickly discovered the advantages of the external adjustable dampers, the new damper design was a great success enjoying wide acceptance from Formula One to all sports cars and prototypes.

Competing in long distance racing in Europe and overseas, especially the USA Koni had already build up a reputation of reliability and success in racing through their importer, Kensington Products Corporation.

Some more advanced damper designs are adjustable but only the rebound force is adjustable and the dampers had to be taken of the car for adjustment, for a road car this is not a problem but for a race car waiting in the pits for the dampers to be taken of the car, dyno tested, adjusted or revalved and installed back on the car takes to much time.

More advanced design twin-tube dampers are adjustable in compression and rebound.

For the compression forces, the valving of the foot valve is made adjustable by increasing or reducing spring load on the compression blow-off valve

The compression damping forces depend on the piston rod diameter; a larger diameter piston rod displaces more fluid than a smaller diameter rod and is therefore more efficient.

10

Flow Damper Design

A variant of the twin-tube design damper is the round flow damper design.

The fluid is always flowing in the same direction, exiting through a passage in the guide, flowing back to the adjuster at the bottom of the damper through a small tube or double tube where it is lead through the adjustable valving system.

The rebound flow as well as the compression flow is directed through the same valving system.

The system is used in inexpensive damper designs, the relation between rebound and compression forces is fixed and can be adjusted simultaneously with one knob.

At the bottom of the cylinder, the system uses a one way valve on top of the blind foot valve and a one way valve on top of the piston

When the damper moves in rebound, the one way piston valve is closed and the fluid above the piston is pressurized between the piston and guide, flowing through a connection in the guide through a fluid line to the valving system where it is met by a spring loaded valve which is adjustable by means of a knob on the outside of the damper body.

When the damper moves in compression, the footvalve does not have a blow-off valve, the fluid below as well as above the piston is pressurized.

The fluid displaced in compression follows the same path through the guide and through the same valving system at the bottom.

During the rebound stroke, fluid is drawn back into the cylinder through the non return valve in the foot valve housing

The difference in damping range between compression and rebound in this design is determined by the relation piston area and piston rod area.

The diameter of cylinder and piston is restricted because there needs to be space for the fluid and air reservoir around the cylinder tube.

The twin-tube damper has to be mounted vertical or nearly vertical because the damper draws fluid back into the working cylinder through the non-return valve in the foot valve, which has to be located at the bottom of the damper and below the fluid level to avoid air entering the cylinder, during the rebound stroke.

The clearance between the piston rod and guide bushing needs to be small 0,03 or 0.04 mm or 0.0025 inch in order to prevent internal hydraulic leaks.

The lip seal is low friction and is capable of withstanding high hydraulic pressure and peak pressure loads when temporary high speed peak velocities are encountered like hitting a curb.

The whole assembly, guide, cylinder and foot valve is pressed down with the top nut, while the cylinder is sealed in some cases with an o ring seal or metal to metal line contact between the guide and the foot valve, so that there are no internal hydraulic fluid leaks.

11

Twin tube damper

The valving consists of orifices for the low velocity damping and a spring loaded valve, which after building up damping force blows of the fluid at mid and higher piston velocities.

The spring loaded valve consists in some cases of a preloaded shim or shims, the preload provided by a nut which can increase or decrease the preload and therefore the damping force, by turning clockwise (increase) or counter clockwise (decrease), the nut is locked with a lock nut so it does not move or loose preload.

In other designs the compression blow-off valve is a flat valve resting on a valve seat, preloaded by a small coil spring or a stack of Belleville washers.

The Belleville washers have the advantage that the stack is compact and the spring rate can be easily changed by stacking the Belleville washers in a different configuration.

During the rebound stroke, fluid is drawn back into the cylinder through a non-return valve on top of the foot valve to fill the volume of the piston rod displaced out of the cylinder during the compression stroke, this non return valve is spring loaded with a low spring force, it closes during the compression stroke to direct the flow through the valving system and generate compression damping force.

The non-return valve on top of the piston opens when the damper begins its compression stroke, to open a passage between the lower part of the cylinder and the top part of the cylinder and closes when the damper starts its rebound stroke so that all fluid is flowing through the valving system underneath the piston.

The outer tube or damper body functions as fluid and air or gas reservoir, the working cylinder is filled completely with fluid, and the reservoir tube for 50 %, when the damper is assembled the piston rod and guide assembly take up space, so that an amount of fluid equal to the volume of the guide and piston rod assembly is displaced out of the cylinder into the reservoir.

The upper half between the cylinder and the outer tube contains the air volume which is necessary to accept the fluid displaced by the piston rod entering the working cylinder the damping forces can be adjustable but in damper designs only in rebound.

Drawing air into the working cylinder results in so called free travel of the piston when the piston travels through air bubbles instead of fluid there is a lack of damping force which is visible on the graph, when the line follows the zero line for a short moment when the piston travels through the air bubble.

In most twin tube damper designs the damping characteristic for the rebound side can be modified by changing the number of bores in the piston or changing the preload on the blow of valve.

In order to make the bores smaller in size, it is necessary to start with a new blank piston which should be available from the manufacturer.

In many cases, the piston bores cannot be changed, because they are powder baked and the fluid passages are standard in the piston.

When an undrilled piston is available I would measure the size and number of bores and calculate the area of each bore.

Multiplied by the number of bores determine the total working area.

Once the area is known, reduce the total area by a certain percentage.

The area of a bore can be calculated with diameter X diameter X ¼ pi gives the area in square mm.

Example 3mm X 3mm X ¼ pi = 3 X 3 X .785 = 7.06 mm square.

VEHICLE DYNAMICS AND DAMPING

6 bores X 7.06 mm square =42.39 total working area so if you want to increase the damping force, select a smaller size bore or reduce the number of bores of the same size.

The easier solution is to keep the same diameter and reduce the number of bores, for example, from 6 to 5 or 4

As a rule of thumb if 6 bores of a certain diameter is 100 %, 5 bores of the same diameter is 83, 3% of the total so the total fluid passage is restricted by 16,6 %.

That way you don't have to go through calculations but you are sure that the rebound force is increased leaving everything else the same.

In the rebound stroke the piston starts its extension travel from the zero point of the stroke, fluid displaced by the piston area starts flowing through the open orifices in the piston rod to the lower half of the cylinder, the lower part of the piston rod is hollow with an oil passage just above the support washer of the piston feeding the orifices.

The larger the piston area, the more fluid displacement per rebound stroke, this is important for a direct acting damper in rebound.

The orifices in most adjustable dampers are drilled for a more accurate adjustment of the damping forces.

A drilled orifice is always of the same size and can be better reproduced so that the amount of fluid flowing through the orifices is the same for all open orifices, necessary to have equal damping forces from one damper to another and make the dampers reproductive in performance.

Some damper designs use a tapered needle and one large orifice to adjust the low velocity damping, changing the amount of fluid passing through the orifice by means of a shaft running through the hollow piston rod, closing or opening a bore in the lower part of the piston rod.

This is accurate when the tapered part of the needle is ground to a perfect angle and the orifice is drilled or honed to a perfect size.

Because it is very difficult to have the correct area between the tapered needle and the bore size from one damper to an other, it manifest itself in the consistency of the damping forces.

The decrease in fluid passage between the tapered needle and the bore diameter Is progressive, when the low velocity damping is adjusted from the zero position to the highest position, the first adjustments from zero do not show a difference in damping force.

The damping force shows the increase from the second half to maximum adjustment, so users adjust the damper starting from the maximum adjustment position to the low adjustment position.

This is the low velocity stage of the extension or rebound stroke, the velocity is too low at that stage and the amount of fluid displaced to small to generate enough pressure to open the blow-off valve.

With increasing velocity the low velocity damping force is building up rapidly.

The resistance the fluid encounters flowing through the open orifices determines the nose angle of the damping graph.

When there are 10 orifices open, the nose angle will be shallow closing the orifices one by one the nose angle will rise to a steeper angle until all are closed and the nose angle will be at its steepest angle.

When dyno tested at a piston velocity of 50 millimetres or 2 inch per second, every time an orifice is closed off, the increase in damping force appears on the dyno graph as a steeper line till all orifices are closed and the line is almost vertical.

This means that the slightest movement of the damper generates a considerable damping force because the maximum low velocity damping is reached within the first millimetres or tenths of an inch of the stroke.

The low velocity damping forces in rebound are so direct in relation to the piston velocity that each time an orifice is closed off it is noticeable in the handling of a car.

When the damper reaches a higher piston velocity, the midrange velocity, the orifices can no longer cope with the amount of fluid flowing through, pressure in the cylinder builds up to the point that the spring loaded blow off valve opens, blowing off some of the fluid flowing through the piston bores.

With increasing velocity to the high velocity range, the amount of fluid passing through the piston bores increases to the point that the piston bores offer resistance, recorded on the graph as an increase in damping force at high velocity, but this happens at a very high piston velocity.

This increase in damping force relative to the high velocity is called the progressivity of the damping characteristic.

I prefer to keep the increase in damping force at high velocity restricted so that the damping force does increase less, relative to the piston velocity.

This will be explained in the chapter about the effects of damping on the handling of a race car.

COMPRESSION STROKE

In the compression stroke of the twin tube damper, fluid is displaced by the piston rod entering the working cylinder.

On top of the piston is a non-return valve, preloaded with a light spring load, this valve lets fluid through only in one way during the compression stroke, to fill the top half of the cylinder and is closed off during the rebound stroke, so that the fluid is forced to pass through the valving system.

On top of the foot valve is a non-return valve which closes during the compression stroke directing the fluid through the valving system of the foot valve.

In the compression stroke, damping forces are generated by the resistance fluid encounters when passing through a combination of orifices and blow-off valve in the foot valve, the amount of fluid displaced by the piston rod is forced to leave the cylinder based on the principle that no two bodies can take up the same space at the same time.

This principle is the basis of the compression damping force, the amount of fluid displaced by the piston rod is equal to the displacement area of the piston rod.

When the damper begins its compression travel, from zero to low velocity, fluid displaced by the piston rod, starts flowing through the orifice or orifices in the foot valve.

The size and the number of open orifices determine the nose angle of the damping graph. With increasing velocity, pressure builds up and the blow off valve opens, releasing fluid into the reservoir showing up on the graph as the point at which the damping force starts to follow a vertical line.

By a good operating blow-off valve, closing well on the valve seat, there can be a dip in the graph line at the point at which the valve opens and the piston speed is still increasing.

This dip is the result of a small pressure drop when the valve opens, fluid is released in a greater quantities than the orifices could handle at the speed it operates at that moment.

With increasing velocity the pressure builds back up again, at this point the spring force on the blow-off valve and the fluid pressure reach equilibrium, the fluid volume which is released and the load on the valve are now in balance so that the damping force line shown on the graph follows a nice horizontal line.

Since the compression damping depends on displacement it follows that a very small diameter piston rod needs a longer stroke to displace the same amount of fluid than a large diameter piston rod.

The objective is to move a race car platform as stable through the air as possible in order to not disturb the airflow under or over the car.

Modern race car suspensions use minimal travel so the dampers have to react to very small movements of the suspension, therefore dampers need to be very effective at low piston velocities.

Modification of the compression damping in a twin tube is mostly restricted to replacement of the foot valve if different valves are available.

Valve seat, valve, flow through area are all fixed and cannot be modified.

In foot valves using a preloaded deflective disc as blow off valve, it is possible to increase the preload on the disc by turning a small screw a ¼ turn clockwise, increasing the damping force.

In order to do this successfully, the screw is secured by a locknut, loosen the nut, without turning the screw.

Turn the screw a quarter turn clockwise and fasten the little locknut without turning the screw or let it turn with the locknut.

Tightening the locknut can add some more preload onto the shims because it takes the play out of the thread this will increase the compression force significantly ...

12

McPherson Strut Design and Application for Sport and Racing Car

The Mc Pherson design simplifies the front suspension of a road car considerably, combining several functions in one component: steering, damping, hub, wheel and brake carrier, spring carrier.

For road cars this development was a big step forward, allowing much more space in the engine bay and a simplified front suspension but the consequence was that the car body needed to be reinforced to take the dynamic loads induced into the car body.

The first application I ever came in contact with was on the Ford Cortina, a British build road car which car was tuned and modified by Colin Chapman and named the Lotus Cortina.

The strut connects the lower control arm direct with the body of the car eliminating the use of the upper control arm and other components saving weight, leave more space in the enginecompartment and more luggage space in case of rear engine cars like the Porsche 911 series.

The centreline of the MC Pherson strut needs to intersect with the centreline of the contact patch of the tire because the strut replaces the kingpin.

In race applications, the centre line does not always intersect but can be a quarter inch away from the centre line of the contact patch, this is called

VEHICLE DYNAMICS AND DAMPING

the scrub radius, the steering is a little heavier but there is a better feel through the steering wheel.

Recently, a team owner told me the story of a modification to the front strut of His race car.

They made a new bracket for the strut (calling it a shock) and put the strut straight up, moving the centre line away from the tire centre line making room for a wider tire.

Because of this modification the scrub radius increased to 10 centimetres (4 inches).

After the modification, they did not understand why the driver could not take corners and the car was slow on the track.

This so called modification destroyed virtually the steering function of the MC Pherson strut.

Every time the car would make a left turn, the front left tire moves to the rear and he right front tire to the front, the steering is extremely heavy and the car unpredictable.

This incident shows that people should have a basic understanding of how a damper works and what the functions are.

We made a Mc Pherson strut application for the Lotus Cortina designed and build by Mr Colin Chapman.

We worked with the Lotus Formula One team of Mr Chapman visiting the design office to discuss new suspension developments with Colin and designer Maurice Philippe and cooperated with tests with the Lotus Formula One team developing a better suspension and handling race car at Silverstone.

During a pause in the test with the Lotus Formula One car at Silverstone, Jimmy Clark drove the Lotus Cortina on the track, performing nice long power slides in the fast corner just before the pit entrance.

One of the best drivers and nicest person on and off the track, during lunchbreak the conversation was about farming and the crops in the field with our importer mr Holland, both farmers.

We were very impressed with the performance and driving skills of Mr Clark, this was the first road car transformed into a race car with Mc Pherson struts known to us, at least in Europe.

Colin Chapman used the strut design also at the rear end of the Lotus Élan, it was a very complicated design, the strut was very difficult to remove from the car, I worked on the development of an insert for the strut but the construction was so complicated that the complete corner of the car had to be removed, not many people would replace the strut so it was not economical to take in production.

For racing applications the front struts had to be shortened in order to fit the lower ride height of the car.

In most early Mc Pherson applications the insert could be easy installed, all brackets, steering arm and wheel hub were welded and bolted on to the strut body so the original housing could be used to upgrade the damping part of the strut.

The damping part of the struts was a twin-tube damper.

Mono tube gas dampers were not suitable for this design because the piston rod of 11 or 12 mm could not take the brake torque, lateral loads induced by the weight, cornering forces etc. into the front end of the car.

The problem for the manufacturers of gas dampers were the high lateral forces, acting on the strut and car body because the weight of the car and the absence of the upper control arm.

In a road car the top-end of the Mc Pherson strut is directly mounted into the car body by means of a strong flexible rubber mount carrying the weight of the car, allowing angular movement of the piston rod, Isolating the body from road noise and vibrations.

The mono-tube gas damper did not have the large diameter piston rod capable of taking the high lateral forces.

This problem was solved by mounting a gas damper upside down into the strut housing so that the damper body performed the function of the piston rod taking up the side loads, brake torque etc.

The flexible rubber top mount performs several different functions, the mount needs to be strong enough to carry the weight of the car, the rubber is needed to isolate the car body from road /tire noise vibrations and harshness, the mount needs to be flexible in order to accept the angular movement of the strut without putting stress on the piston rod when the front suspension runs through its travel.

In some mc Pherson strut designs the guide in which the piston rod rides was made flexible, the guide could accept angular movement to some degree, preventing the damper part from causing too much internal friction.

The movable guide design eliminated any friction or stiction as it was called and offered better ride qualities for road cars.

The rubber in the mount needs a lot of fine tuning by the engineering department of the car manufacturer, the weight of the car rests on the mount, preloads the rubber to the extent that it effects the isolation function and one of the first requirements of the rubber mount is to carry the weight of the car but at the same time isolate road noise and vibration from the car body.

Preloaded rubber transmits more vibration and noise when the rubber compound is of certain shore hardness.

When the rubber compound is to hard it will transmit road noise and vibrations easily into the car body, If the rubber is to soft, the mount will be to compliant and make the car directional and in cornering unstable because the top of the mc Pherson can move laterally.

In race applications, the rubber mount is replaced by a heavy spherical bearing, allowing angular movement, performing the same functions as the rubber mount except that there is no flexibility but suspension movements are transmitted directly without the compliance of the rubber mount and of course with all the noise and vibration going on in a race car the isolation function can be neglected for this application.

The top of the damper towers in which the struts are mounted can be reinforced with tower braces, connecting the two strut mounting towers across the engine bay with a bar bolted onto the damper mounting bolts, the bar prevents the two damper towers from flexing laterally so that the directional stability and cornering is improved.

This kind of cross bar is available from tuning companies.

The use of mc Pherson struts requires strong bodywork in the front end of the car capable of taking the forces induced by lateral loads because of the weight of the car, brake torque, acceleration torque by front wheel drive and high lateral force due to cornering.

The whole car body and interior acts as a large amplifier, transmitting noise and vibration into the car's interior which is very annoying for passengers in a road car.

In the early design twin-tube dampers, a powder baked piston or cast iron piston was used but the metal to metal contact between the piston and cylinder of the early mc Pherson strut designs caused so much internal friction that it had a negative effect on the ride quality of the cars.

The friction was so severe that when the front of the car was pushed down by hand, the car stayed down and the spring force could not overcome the friction and the static weight of the car, as a result the car body did not return to its static ride height as it should have done without the friction.

Friction is a breakaway force causing a harsh and jarring ride over smaller bumps.

The solution was to use Teflon coated guide bushing and Teflon coat the piston or use a low friction Teflon band around the piston in some cases.

To eliminate the friction the outside diameter of the piston was grooved and a Teflon washer was forced over the material by pulling the piston through a mandrel under pressure and heat.

The Teflon flows, filling the grooves and after cooling down, stays in place during operation of the damper while the outside diameter is within small tolerances for a perfect fit inside the cylinder.

The result was a perfect fitting piston without causing any friction when lateral loaded.

For the Mc Pherson strut it made a difference like night and day, the suspension would move freely when the car was pushed down by hand and would come back up again to ride height, while the friction was reduced to almost zero and the ride quality improved considerably.

The mono tube gas damper used a 12 mm piston rod because the single tube design needed 25 bars of gas pressure in order to make the gas damper function in compression.

Using a large enough diameter piston rod capable of taking the lateral forces in combination with the high gas pressure was not believed possible, until we just made one proving everybody wrong who said that it could not be done.

The 22 millimetre piston rod we use displaces more fluid and there is enough space in the reservoir to accommodate the divider piston and sufficient gas volume for the strut to function well.

A point of attention is that the brackets are welded direct onto the strut housing causing some distortion of the cylinder tube and therefore the strut housing has to be honed after welding to be perfectly round.

In some other designs the lower bracket is clamped onto the strut body, the clamping force can jam the divider piston so the strut body has to have a certain wall thickness to prevent the divider piston from jamming in the strut body.

In general it was believed that the gas pressure needed to make the single tube damper design work in a Mc Pherson design would provide too much lift, affecting the ride height which has not been a problem in anyway.

The disadvantage of putting a gas damper inside a strut housing is that the damper riding inside the strut housing requires Teflon guides to avoid metal to metal contact between the damper body and strut housing.

The Teflon guide bushings are dry, not lubricated with grease; Teflon guides reduce road noise providing low friction contact.

In road and rally cars using more suspension travel, the damper body works as an air pump displacing large amounts of air in and out of the strut housing any time it moves, drawing road dirt like sand, mud and water into the housing during the extension stroke.

The volume of air displaced in this system needs to move unrestricted from inside the damper housing to the outside, dirt entering the strut housing, wears out the Teflon guides quicker than a closed damper design, causing an increasing louder noise when used in road cars.

The twintube Mc Pherson strut was designed with a larger diameter piston rod of 22 mm or 25 mm in some cases capable of withstanding the lateral loads using a gas pressure of 6 bars.

The 6 bar of gas pressure produced a positive lift of 22.74 kg, 50 or pounds and 29.4 kg or 65 pounds respectively.

The twin tube damper does not depend on gas pressure to generate compression forces but compression damping is generated through the foot valve so the 6 bar of gas pressure works just fine.

The twin-tube damper in the Mc Pherson strut design remained the same in design and operation but the gas pressure supports part of the weight of the car and keeps the tire better in contact with the road improving the ride quality over bumps and isolating the car body better from irregularities and road noise.

The gas pressure in the twin tube damper requires an extra one way valve and passage in the top of the cylinder to remove any gas that could have entered the working cylinder and mixed with the fluid during operation.

Overnight, when the car is parked, gas diffused (absorbed) into the fluid will accumulate in the top of the cylinder of strut or damper as a gas bubble.

For this condition, the guide of this design damper provides an escape passage for the gas, there is an orifice drilled in the guide opening providing a passage from the cylinder to the reservoir.

The gas bubble will be forced out of the cylinder flowing through the oil/gas passage and the one way valve as soon as the strut or damper starts moving.

The valve is a rubber lip seal, letting gas through out of the cylinder but is closed for the return flow back into the cylinder.

By absence of an escape valve, the gas bubble is likely to produce a knuckling sound on cold mornings in road cars, referred to as morning sickness by the car engineers.

Any time the piston moves through a gas bubble, the damper will produce a hydraulic knock which is experienced as a knuckling sound in the car.

Air or gas inside the working cylinder effects the damping performance in a negative way, when the piston encounters an air or gas bubble, the flow of fluid is interrupted resulting in free travel of the piston.

I have done extensive development work with the low pressure gas strut design and it was a positive development in terms of ride and handling of road cars, more comfort with the same or better handling characteristic

The first strut designs had the wheel hub carrier welded or pressed onto the Mc Pherson strut housing.

Porsche used a forged wheel carrier for the early 911 complete with steering arm and a bracket for the brake caliper.

So when the damper function is not up to date anymore, the whole strut has to be replaced, which is not very practical and a waste of good parts but the used wheel carrier can be removed and used to build a new replacement.

The outside cylinder was rolled close at the top but when some of the material was machined away the internals could be removed, the strut body was already threaded on the inside so that a replacement strut could be installed using the existing thread to tighten down the cylinder.

Aftermarket suppliers used those threads to install inserts so that the damper function was restored and improved because the insert was usually a better quality damper and in some cases adjustable in rebound than the original equipment struts.

The adjustability of the early inserts was not very effective the strut had to be removed from the car to adjust, compressed, the piston rod turned and installed again

The road application used a welded on spring platform for the spring to rest on, in some cases, the strut body is grooved so that the lower spring platform can be adjusted up and down in increments of 10 mm.

The fixed spring platforms were a stamped product, formed so that the end of the spring would fit into the platform.

The end of the spring would fit into the contoured spring platform preventing it from turning so that the spring helped to keep the steering wheel return to the straight forward position.

For race applications the strut body was machined and threaded, the platforms are slotted and can be locked by means of a bolt clamping the platform around the strut body or use a locking nut underneath the spring platform.

Those spring platforms can be adjusted infinitely, making it easier to adjust the cross weight of a car exactly on the scales.

The original springs were replaced by cylindrical springs instead of the tapered form.

The cylindrical springs are smaller in diameter and allow more space for wider tires used in racing applications

A bracket is welded on to the strut body, so that the wheel carrier can be bolted on to the bracket to use a greater offset of the hub centre so that there is more room for wider tires.

The greater offset will increase the scrub radius of the tire but this does not seem to be a problem, for some applications, the brackets are equipped with slotted bolt holes to provide camber adjustment, the slotted holes are machined out to accept a washer with an eccentric hole for the bolt to fit through.

The washer fits into a recess machined into the bracket so the outside diameter keeps the camber setting fixed moving the strut housing to or away from the wheel spindle, turning the washer changes the camber setting to more or less negative camber on the front as well at the rear of the car when the rear axle is equipped with struts.

The camber change is very effective but more difficult to execute because the bolts holding the hub have to be loosened and it is difficult to hold the bracket in place with the bolts.

Most other designs use a top mounting plate with slotted holes so that the camber can be adjusted form the top although the top of the strut has to move quit a distance to make a significant camber change.

It is easier to execute than the bracket adjustment and in most cases the change is large enough to make considerable camber adjustments

Using this top plate adjustment design, the wheels can be left on the car making it easier to measure and change camber and toe settings.

13

THE DAMPING FUNCTION OF THE TWIN TUBE STRUT/DAMPER

The velocity ratio of a Mc Pherson strut is in most cases 1:1.

This is positive for the damper function of the strut because 1 inch movement of the wheel is equal to 1 inch movement of the strut or close to that.

Dampers, whether they are mounted in the suspension as a damper or in a strut, need travel to function, without displacement and velocity the strut or damper is a dead object except for the gas pressure exerting a force on the piston rod permanently.

Suspension travel means displacement of fluid inside the damper, the performance of any damper depends on displacement, valving design and piston velocity.

The sturdy piston rod of the Mc Pherson, 22 mm or in some cases 25 mm, displaces a large amount of fluid through the valving of the foot valve.

Therefore the compression valving in a Mc Pherson strut has to be designed for a large flow of fluid, building up enough damping force to control suspension travel

Valve seat diameter, valve, spring load have to be in balance with displacement, in compression, the fluid is forced out of the cylinder through the foot valve from the zero point of the suspension travel, fluid flows through the orifices of the foot valve.

With increasing velocity the orifices cannot accommodate the amount of fluid forcing the blow off valve to open and all fluid displaced by the piston rod flows out of the cylinder into the reservoir which is the space between the cylinder and the outer tube of the strut.

Reservoir space is restricted in some designs because in race applications there is always the tendency to keep packaging sizes as small as possible

When filling a Mc Pherson strut with fluid, the cylinder should be filled completely without overflowing because the Mc Pherson strut designs requires a certain distance between the guide and the piston, called the bearing span so that the high lateral forces are divided over a certain length.

When the piston and rod assembly is mounted inside the strut, fluid will be displaced by the piston rod /piston assembly, so much fluid is displaced that there is plenty of fluid in the reservoir for a good operation of the strut /damper after assembly.

REBOUND DAMPING TWIN TUBE STRUT.

The rebound damping in a twin tube Mc Pherson strut works the same way as the twin tube damper.

During the rebound stroke of the Mc Pherson strut, fluid in the cylinder above the piston is compressed and flows through orifices from the top of the cylinder to the chamber below the piston, this is the low velocity damping.

With increasing velocity, the flow of damper fluid becomes too much for the orifices to handle and pressure builds up against the blow-off valve.

When the pressure overcomes the spring load of the valve, the blow-off valve opens and the fluid flows to the lower part of the cylinder.

Sometimes the blow off valve is adjustable by increasing the spring load onto the valve while at the same time orifices can be closed off so that the nose angle of the graph as well as the high velocity damping can be adjusted.

A longer rebound stop is required so that there is a certain distance between guide and piston at full droop to avoid large lateral bending forces at the piston rod.

14

McPherson Struts For Race Applications

Designing mc Person Struts for race applications is a challenge.

The design needs to be light weight to keep the unsprung weight as lo as possible but very strong at the same time to take the various loads induced into the strut housing and piston rod.

The Mc Pherson strut for race applications is a single tube design damper with external reservoir and four ways adjustable for the top level of racing series.

For club and road applications the strut is available as a twintube damper, two way adjustable.

Because of the lower ride height the struts need to be shorter presenting a problem with the fluid and gas volume.

Race cars like the Porsche 993 996 997 etc. equipped with front struts running on tracks like Daytona or Homestead where they run part of the speedway oval, produce a lot of vibration through the suspension, the vibration can be so severe that the top nut of the strut shakes loose, the nut has to be secured with safety wire or a setscrew to prevent the nut from coming loose.

This would not cause any problem for the function of the damper but when the nut comes loose it can scrape the chrome plated surface of the piston rod, damaging the finish of the rod which in turn can damage the seal starting a fluid leak and if that continues for a longer time, the front strut has to be replaced duriong the race which is bad news for a race car.

I have been designing struts for racing applications for a long time, the demands put on the design for Mc Pherson struts for racing cars are much more severe than on road car applications.

The offset of the wide racing tires with optimal grip put high lateral loads into the strut when cornering and certainly when the race setup produces optimal grip and traction the strut is subjected too much abuse.

Special precautions have to be taken when designing a Mc Pherson strut for racing applications.

The piston rod of the Mc Pherson strut has to be strong enough to withstand the lateral load without bending or breaking under stress, whereas the piston rod of a damper which is mounted between spherical bearings can be of carbon steel hardened and chrome plated, the material for a race Mc Pherson strut has to be of a different, stronger material, another requirement is the distance between the guide and the top of the piston, sometime called the bearing span.

If the distance is to short, the whole front suspension assembly can move lateral because of the necessary tolerances needed for the moving parts like piston and piston rod.

It is not a real big problem but people perceive it as a problem, the only time the play is noticeable is when the car is jacked up and the wheels are of the ground.

As soon as the car is back on its wheels again, the suspension is loaded and any play is pushed to one side and the car or driver would not notice any difference.

The requirements for building a Mc Pherson strut for racing applications are different than for dampers while the conditions are more demanding.

Design and the construction of the piston assembly and guide are very important for the life time and an optimal performing front suspension.

Besides using a stronger material, treatment of the material is also very important for the life time of the racing strut,

The material needs to be induction hardened/water quenched and ground to a precise diameter before chrome plating and super finished after chrome plating.

The chrome layer is about 0,017 to 0,025 millimetres thick, the chrome plated rods need to be stress relieved right after plating to prevent hydrogen brittleness.

Hydrogen brittleness is a condition of the material right after it comes out of the chrome plating bath, which, if not treated is cause for the material to break.

After heat treatment, the material is elastic like a spring and can bend to a certain degree without taking a permanent set; this is one of the tests to check the strength of the piston rod.

The dynamic loads induced into the Mc Pherson struts originate from several outside sources.

First there is the static weight of the car pushing the wheels to the outside, ad to this the dynamic weight of down force, lateral cornering forces at high speed, taking corners under braking and we have a variety of causes threatening the life cycle of the Mc Pherson strut.

In a 24 hour race, the last thing you want to happen is having to replace front struts, the car has been balanced, cambers and casters set, corner weights adjusted so we build our racing struts to last many 24 hour races without repairs or revalving.

15

Single/Mono tube damper design and operation

The single tube or monotube damper is named that way because the working cylinder, fluid and gas reservoir are situated in line (axial), the body of the mono-tube damper functions as cylinder and damper body in one.

The early monotube gas dampers were designed by a company "De Carbon" in France and the damper was called "system de carbon" later in time adopted by other companies making the gas pressure damper widely available to many more applications.

The piston of the mono-tube rides direct inside the damper body/cylinder, metal to metal contact between piston and cylinder wall is prevented by a Teflon band which is located in a groove in the piston supported by an o ring preloading the Teflon band against the cylinder wall eliminating any friction and fluid leaking, bypassing the piston valving.

The guide is equipped with a Teflon coated bushing preventing metal to metal contact between piston rod and guide.

In the single tube damper, the piston is bolted onto the piston rod with the bump or compression valving shim stack at the topside of the piston and the rebound valving shim stack located on the lower part of the piston.

Valving stacks are building up from steel shims with spring steel quality, the largest diameter covering the piston bores, continuing with smaller diameter shims, thickness of the shim is selected by the damper engineer depending on damping forces and characteristic required.

The deflection of a shim blowing off fluid is minimal, not more than a 0,2 millimetre or 0,008 inch at low to moderate piston velocity, at higher velocities the shim might open a little more.

At high piston velocity the shim will open wider to let more fluid pass through.

The diameter of the shim covering the piston bores need to be as large as possible, the shim is clamped in the mid-section onto the piston leaving the outer diameter to flex, letting fluid pass through the piston bores controlling the damping force by the resistance of the flexed disc.

When the valve opens, the piston velocity increases at the same time so that the damping force remains constant, in case a digressive damping characteristic is needed.

The fluid Column is separated from the gas volume by a floating divider piston, sealed on the outside diameter by a large O ring seal, a two way lip seal or quad ring seal, in some cases there is a second groove holding a Teflon band to limit friction and keep the divider piston from jamming in the cylinder.

The piston rod diameter is in general 12 millimetre 4,7 inch or 14 millimetre, 0.55 inch

The displacement of a larger diameter rod causes an increase in gas pressure in compression.

The monotube design restricts the packaging length of the damper and is at a disadvantage in a suspension requiring short damper lengths.

A lowered suspension requiring a short packaging length or very low open wheel race cars.

In this design, limiting the piston rod diameter is necessary because of the small gas volume available to accept the fluid displaced by the piston rod

A large displacement of damper fluid causes the gas pressure to increase too much, while the gas volume is restricted by the length and diameter of the damper.

The gas volume needs to be large enough to absorb the amount of fluid displaced by the piston rod so that the pressure increase is kept at a moderate level when the damper is compressed.

Those factors are limitations in the monotone design not found by the twin tube damper design or the monotone with external reservoir.

In order to generate sufficient higher compression damping forces, the gas pressure has to be 25 bars.

The back pressure is to prevent the divider piston to be driven into the gas reservoir by very high compression damping forces.

In the mono-tube damper design the compression damping forces are generated by the shim stack on top of the piston, the largest shim functions as a blow-off valve.

At a very low piston velocity, the blow-off valve is closed and fluid flows through the orifices before the piston velocity is high enough to open the valve stack (blow-off) valve.

The gas pressure is providing back pressure to prevent the fluid Colom from being driven into the gas reservoir.

The maximum compression damping force is limited by the back pressure of the gas volume, so the maximum compression damping force depends on the gas pressure in the gas reservoir.

The gas pressure works against the cross section of the piston rod, which by a diameter of 12 millimetres is approximately 1 square centimetre x 25 bar pressure delivers an extension force of 25 KG or 55 pounds of lift supporting the suspension.

In the past, people have asked how much the ride height of a car would be affected by the gas pressure of a mono-tube gas damper.

The answer is, very little, because in case of a road car 44 pounds of lift is such a small percentage of the total weight of a car that nobody will notice a difference in ride height certainly not the dynamic ride height.

The ride height on racing cars is adjustable so there is no problem at all.

The advantage of this type of damper is that it can be mounted in any position, horizontal, upside down or under an angle.

Special the fact that they can be mounted upside down is attractive to race engineers, the manufacturers claim that it lowers the un-sprung weight of a car, which is debatable since the steel piston rod assembly usual weighs more than the aluminium cylinder with fluid and when the damper is mounted upside down, the piston rod assembly is part of the un-sprung weight of the car.

A drawback of the monotube design is that when the damper body gets dented by gravel or rocks, the piston jams and the damper is inoperable, this might not happen often at the race track but in rallying it might happen and is a setback.

In a mono-tube damper design there is only one tube which is used as damper body and cylinder, hence the name mono tube damper.

Operation

Working cylinder and gas reservoir are placed in line, the cylinder at the top part and the gas volume at the bottom end of the damper, fluid and gas separated by a divider piston

The top side of the cylinder is closed with a guide, the piston rod rides in a Teflon coated bushing, the guide is secured with a circlip lodged in a radius, machined in the inside wall of the cylinder and a radius on the outside top of the guide.

The guide, sealed on the outside with an O ring and the inside with a lip seal around the piston rod is kept in place by the gas/oil pressure, with a gas pressure of 25 bars, a damper with a 44 mm inside diameter is held in place with a force of 355 kg or 780 pounds.

This is a considerable force which can be dangerous when a gas pressurized damper is opened by inexperienced people, the gas pressure needs to be released before the damper can be opened safely but in some designs the damper is sealed and special tooling is required.

When the dampers are equipped with Schrader valves, the gas pressure can simply be released by taking out the internal part of the valve.

The divider piston is sealed on the outside diameter with an O ring seal or a quad ring seal and in some designs there is an extra Teflon band to ease friction when the piston moves through the cylinder.

The piston rod /piston assembly, is located in the upper part of the cylinder

The cylinder is filled completely with damper fluid, while the divider piston acts as the bottom for the fluid Colom, the rebound valving (valve stack) is located at the top of the piston and the compression stack underneath the piston, secured with a nut.

Rebound stroke

During the rebound stroke, the piston rod is extracted from the cylinder, trapping damper fluid between the piston and the guide, with pressure building up at low velocity, fluid starts flowing through one or more orifices, with increasing velocity, pressure builds up and the valve opens, releasing the fluid into the lower part of the cylinder, the divider piston takes up the volume vacated by the extending piston rod moving with the fluid column.

The gas pressure underneath the divider piston keeps the fluid Column pressurized forcing the divider piston to follow the movement of the piston rod/piston assembly.

The rebound damping characteristic depends on the configuration of the valve stack; a stack with many shims stacked up with small differences in diameter will result in a progressive characteristic, meaning that the damping force increases faster relative to the piston rod velocity.

For a digressive damping characteristic the shim stack has to be changed, instead of many shims decreasing in diameter in small measurements, the digressive shim stack can be built with a small number of shims but with more preload or thicker shims so that when the pressure opens the valve (shim stack) more fluid is released in a short time frame.

The valve stack will release so much fluid that the increase in piston velocity does not have an effect on the damping characteristic and the damping force remains at the same level with increasing piston velocity, which is a digressive characteristic.

The bores in the piston are fixed in damper designs but some manufacturers believe in using many different configurations, the most intricate figures are sometimes milled into the piston surface as valve seat but the basic function of providing resistance to the fluid pressure building up as a result of the moving piston remains the same.

The total area working on the blow-off valve determines when the valve will open, not the configuration of the area.

Compression (bump) stroke.

The compression or bump damping forces are generated by displacement of the piston rod fluid volume when the piston rod enters the cylinder in the compression stroke, displacing the volume of the piston rod driving the divider piston deeper into the gas volume.

At low piston velocity damper fluid flows through orifices into the upper part of the cylinder and with increasing velocity the hydraulic pressure opens the valve, releasing the fluid into the upper part of the cylinder.

This may sound a bit confusing to the non-damper technician because the piston rod entering the cylinder displaces fluid which in turn displaces the divider piston further into the gas volume while at the same time fluid flows from the lower part of the cylinder to the upper part of the cylinder.

The reason is that the space of the moving piston area has to be filled and at low velocity an orifice can coop with the flow of fluid but with increasing velocity the amount of fluid displaced is larger and the valves on top of the piston need to open letting fluid pass to the upper part of the cylinder.

If the valve would not open the piston pushes the fluid into the reservoir and the space above the piston would be empty, filled with air.

This could be the case when the compression forces are extreme so that the gas pressure is not able to provide enough back pressure.

One has to consider two things here, when the piston moves in compression, the piston displacement of a 44 mm piston diameter is 15 cm square so as long as the compression valving remains closed, damping is generated by the piston area of 15 square centimetres against the back pressure of the gas pressure.

As soon as the valving releases fluid, damping is generated by the volume displaced by the piston rod which is a much smaller area (1 cm square) in most designs, but because the piston velocity has increased so much the damping force stays at the same value with the valve open and will decrease when the compression stroke slows down to the zero velocity point, starting to move in the rebound stroke again.

In this design the gas pressure exerts an extension force on the piston rod of 25 bar equalling 25 Kg or 55 pounds of force, this force is always present whether the damper is installed in a suspension or whether the damper is tested in a dyno.

When running a gas pressurized damper in a dyno, the gas force should be zeroed so that the real damping forces are recorded on the graph, sometimes people forget that and get two different graphs on the same paper which can be confusing and lead to wrong conclusions when not detected.

Several times in my career I have been confronted with graphs showing two different values with the question why there was so much difference between two or more dampers.

Taking a closer look, the only thing wrong was that, testing one damper the gas force was zeroed while the other damper was tested without zeroing the gas force.

In a damper with a small piston rod diameter the difference is not that much, with a pressure of 20 bar the force exerted on the piston rod is 20 kg or 44 pounds.

However when the damper is equipped with a 22 mm piston rod, 20 bar of pressure delivers a force of 75,8 kg or 166,75 pounds, a considerable amount of force and when two dampers are recorded on one graph and of one damper the gas force is zeroed while the other damper is not, the graphs are in different places while they are supposed to be overlaid.

The result is that it seems that the dampers are different in force because the damper graph lines are in different places on the graph sheet but when the graphs are overlaid the forces should be exactly the same.

JAN ZUIJDIJK

Most mono-tube gas dampers are nonadjustable, some manufacturers offer external adjustable mono-tube dampers.

When a mono-tube is selected for a race application, I would recommend to take an external adjustable mono-tube damper, a nonadjustable damper has either to be revalved or replaced with another damper with different valving, when the track conditions call for more or less damping, which means that several sets of dampers have to be carried around to the race track in order to select the right valving for that particular track.

16

Mono Tube Damper With External Gas/Fliuid Reservoir

The use of external fluid/gas reservoirs have often been criticized by competing manufacturer claiming that they have the better idea, not having to use a reservoir connected to the damper or strut body by a pressure hose.

What they fail to say is that the gas reservoir in mono-tube design dampers without external reservoir space can be restricted to the point that the gas pressure increase by completely compressed struts or dampers is so much that the small gas volume shows a sharp increase in pressure.

This sharp increase in pressure ads a certain harshness to the ride of the car, it feels like a progressive bump rubber is being compressed or a stiff helper spring is used.

The handling of a car utilizing such a design can also be affected and is difficult to tune because of the small gas volume.

The small gas volume, when compressed, acts like a very progressive spring or bump rubber, adding to much force to the compression end of the stroke.

I have designed a mono-tube gas damper with an external gas/fluid reservoir for the Indy car of Patrick Racing driven by Mario Andretti.

We tested the dampers during practice of the last Indy car race at the Riverside race track in California, the improvement in handling of the car was so great that Mario could not believe that the ride could be so smooth and handle well; as a matter of fact the car was seconds faster with the new design dampers.

For some damper designs it is necessary to increase the fluid/gas reservoir volume outside the damper body, the outside diameter of the damper body does not always provide enough space for the fluid volume which is displaced out of the working cylinder on the compression stroke.

In this situation the designer can do one of two things, reduce displacement by using a small diameter piston rod so that the fluid /gas reservoir can accept the fluid displacement without increasing the gas pressure to much.

The better way, is using an external reservoir, based on the theory that dampers in modern race car suspensions need as much fluid displacement as possible per centimetre stroke, In order to provide effective damping performance at very low piston velocities.

In the early days of Mc Pherson applications the strut housing was designed to function as fluid reservoir, in most designs the wheel carrier bracket was pressed onto the damper body tube and welded on, the press fit offered already a strong connection but the weld added extra security against the assembly coming apart.

In order to fit the bracket the reservoir tube was narrower than the normal damper body, the smaller diameter tube restricted the reservoir space for the twin tube damper inside the strut so the damper function was not optimal.

With the development of the Porsche 911, the brackets where pressed on to the strut body with a force no less than 9000 kg or 19800 pounds force.

In addition the bracket had to be welded on the backside, all this to prevent the bracket from moving on strut the body and coming loose.

It was overkill but better safe than sorry.

Special in some McPherson applications, the McPherson strut housing (damper body) is too small in diameter to offer enough reservoir space to absorb the fluid displaced out of the cylinder during the compression stroke.

VEHICLE DYNAMICS AND DAMPING

Mc Pherson struts combine several functions, steering, wheel carrier, spring carrier, taking up lateral forces, brake torque, acceleration forces put into the strut require a strong piston rod but this means also a large amount of fluid displacement In addition, the gas reservoir needs to have an extra volume available so that when the fluid volume which is displaced into the reservoir, there needs to be ample space to avoid a large increase in gas pressure when the damper is compressed.

As a rule of thumb, the gas reservoir space needs to be minimal 3 times the displaced fluid volume in order to keep the pressure increase at an acceptable level when the damper is completely compressed so a strut with a 22 millimetre diameter piston rod and a 120 millimetre stroke, displaces 45, 48 Cubic centimetre fluid, the available reservoir space needs to be 160. Cubic centimetre.

For an accurate computation of the gas pressure increase when the damper is compressed, Boyles law applies, Pressure X Volume = Constant.

Boyle's law says that, when you have a reservoir with a gas volume of 160 cc gas under a pressure of 20 bars, the constant is 160 X 20 = 3,200

When the piston rod displaces 40 cc into the reservoir, the gas volume is reduced from 160 - 40=120 cubic centimetre ...

The gas pressure will then be 3,200 divided by 120 = 26,6 bar so the gas pressure is increased with 6.6 bar over the total stroke but because the damper is only compressed completely in extreme cases and works at the midpoint of the stroke the average pressure increase will be 3,3 bar

This exercise is done without taking into consideration the increase in temperature which has a certain effect on the pressure increase also but that would complicate matters more than necessary, the idea is that the air or gas volume should be large enough to keep the pressure increase at a minimum during operation of the damper

The external reservoir is used as an expansion vessel, accepting the amount of fluid displaced out of the cylinder on the compression stroke returning the fluid to the damper on the rebound stroke.

The external reservoir is also useful to cool the fluid of struts or dampers used under very straining conditions like rally cars and race cars on very bumpy race tracks, for the rally and off road applications, some manufacturers use cooling fins on the reservoir.

The external reservoir can also be connected to the damper or strut body by a high pressure hose so that the reservoir can be placed in a location out of the way in the engine compartment or in the trunk.

Another possible construction is to connect the reservoir direct to the damper body, so called piggy back design in which case the reservoir is connected onto the damper body with tubing and clamped on to the damper body.

The application of an external reservoir without valving means that the heat generation In this design takes place inside the damper body, in the twin tube damper, the damping forces are generated in the piston(rebound) and the foot valve (compression).

In the mono-tube design damper, heat is generated in the piston valving in both rebound and compression damping forces.

Using an external reservoir just as an expansion vessel does not change the principal operation of the mono-tube damper function.

The advantage is that there is plenty of space to accept the fluid volume displaced by the piston rod without increasing the gas pressure in the reservoir so that there is no change in damper performance when the damper temperatures increase.

17

Low Pressure Twin Tube Gas Damper

With the mono tube gas damper gaining in popularity, it was necessary to develop a successful combination of the twin tube damper and mono-tube damper since the existing twin tube gas damper was not designed to hold a gas volume under pressure.

The manufacturers of mono tube gas dampers and the twin tube hydraulic damper manufacturers, both believed that their concept and design was the best solution for the suspension of the auto industry, but the mono tube gas damper although not adjustable, gained more terrain, so the twin tube hydraulic damper manufacturers had to react and find a suitable alternative.

Gas seemed to be a magic word in the industry while in the racing world, gas dampers were not widely used in competition yet but for road cars, the gas dampers provided a nicer what is called boulevard ride.

The science of using the damping characteristic for tuning purposes was not very much explored in the those days except at koni where the damping forces were adjustable from the outside (external adjustable).

Using the nonadjustable single tube gas damper for racing applications, teams had to carry several sets with different damper characteristic

The advantage and simplicity of the monotube gas damper over the twin tube hydraulic damper was evident, although the developments as we know now were a long way off.

For this reason the low pressure twin tube damper was invented; the damper, a twin tube damper design in which the atmospheric pressure of the air volume in the reservoir was replaced by 6 bars about 85 PSI of gas pressure, which was considered low pressure, in comparison with the 25 bar of pressure used in the mono-tube tube gas damper.

To make the low pressure gas damper work, some design modifications of the twin tube damper were necessary.

Because the gas volume was located in the reservoir space in-between the working cylinder and the outer tube there was a real possibility that gas would enter the working cylinder through the space between the piston rod and the guide.

In most twin tube dampers, air and some fluid are transported out of the working cylinder during the rebound stroke, passing through bores in the guide into the reservoir space.

When the reservoir space is charged with 6 bar of gas pressure, gas can take the same route flowing back into the cylinder were it would form a gas bubble in the damper fluid at the top of the cylinder.

To avoid this condition from happening a one way rubber valve is installed allowing the gas and fluid passing through the bores only out of the cylinder through the valve but cannot return.

The one way valve consists of a rubber seal which is located around the lower part on the outside of the guide, sealing the outside of the reservoir tube permanently and closes around the guide with a lip seal.

The lip seal allows fluid and gas to pass from the cylinder into the reservoir but the return flow is blocked by the one lip seal.

Any gas accumulated in the top of the cylinder will leave the cylinder at the first movement of the suspension through the escape valve.

The 6 bars of gas pressure exerts a force in extension (rebound) of 6 kg or 13,5 pounds by a 12 mm piston rod.

When using a 16 mm piston rod, the force extending the piston rod is 12 kg or 26 pounds.

The lifting force is not really affecting the ride height of a car since the force is too low a percentage of the total lift, provided by the springs, to worry about.

The ride quality of a car equipped with low pressure twin tube gas dampers is much better than the twin tube damper with atmospheric pressure; the suspension travel is better controlled and dampened.

The comfort level of road cars using the low pressure gas system is a lot better while the handling remains the same or is better than cars equipped with normal twin tube hydraulic dampers.

We have worked on the development of this system in the early days of its existence and the results in ride and handling are remarkable, the car rides the bumps on the road much better, transmitting less road noise.

Special the Mc Pherson strut, at first used only in the front of a car but in some later car models also at the rear of the chassis benefits from the lift and the preload on the suspension of the gas pressure.

The Mc Pherson strut combines several functions.

The strut connects the lower control arm direct with the body of the car making the use of the upper control arm unnecessary, saving space in the engine compartment and more luggage space in case of rear engine cars like the Porsche 911 series.

The problem for the manufacturers of gas dampers were the high lateral forces acting on the strut and car body because the weight of the car and cornering forces.

The top end of the Mc Pherson strut is directly mounted into the car body, putting high loads into the front end of the car body so that the front end of the body had to be reinforced while this system needs more attention to the top mounts isolating the car body from road noise and vibration.

For twin tube dampers a powder baked piston or a cast iron piston is used, the metal to metal contact between piston and cylinder of the early Mc Pherson struts caused so much friction that it had a negative effect on the ride quality of the cars.

The friction was so severe that when the front of the car was pushed down by hand, the car stayed down and the spring force could not overcome the force of the friction, the car body did not come back up back to its static ride height as it should have done.

The solution was to Teflon coated the piston using a Teflon band around the piston, Teflon is a low friction material.

That made a difference like night and day, the suspension would move freely and the friction was reduced to almost zero.

Friction is a breakaway force, it means that the damper will move only after a higher force is applied than the friction resisting movement, feeding all jolts and jerks direct into the car body making it a very uncomfortable ride for driver and passenger.

The mono tube design damper needs 25 bars of gas pressure to make it function in compression; using a large enough diameter piston rod to withstand the lateral forces in combination with the high gas pressure was not believed possible.

So the gas damper manufacturers found the solution in installing a normal gas damper upside down inside the Mc Pherson strut housing using the damper body to take up the lateral load of the Mc Pherson strut.

The disadvantage of this design is that the damperbody functions as pistonrod riding inside the strut housing, by cars using more suspension travel, the damper body displaces a large amount of air in and out of the strut housing any time it moves, working like an airpump, drawing road dirt like sand mud and water into the strut housing, wearing out the guides quicker than in a closed damper design.

The twin tube design Mc Pherson strut was equipped with a larger diameter piston rod of 22 mm or 25 millimetres, able to withstand the lateral forces, using a gas pressure of 6 bar.

The 6 bar of gas pressure produced a positive lift of 22.74 and 29.4 kg 50 and 65pounds respectively, it is hard to say where the 6 bar of gas pressure came from or who decided to use that figure but it was generally accepted as a good starting point.

The compression damping forces in the twin tube design do not depend on gas pressure to function but compression damping is generated by displacing damper fluid through the foot valve so the 6 bar of gas works just fine.

The twin tube damper remained the same in design and operation but the gas pressure helped to carry part of the weight of the car, improving the ride quality for road cars over bumps and isolating the car body better from road irregularities and road noise

The gas pressure in the twin tube damper requires an extra one way valve in the top of the damper to remove any gas that would have entered the working cylinder and mixed in with the fluid during operation.

Overnight when the car is parked, gas which is diffused into the damper fluid will accumulate in the top of the cylinder of the damper, the gas bubble will escape through the valve as soon as the strut or damper starts moving.

When there would not be an escape valve, the gas bubble inside the cylinder is likely to produce noise on cold mornings in road cars, called morning sickness.

Any time the piston moves through a gas bubble, the damper will produce a hydraulic knock which is experienced as a loud knuckling sound in the car.

I have done extensive development work with this design and it was a positive development in terms of ride and handling of road cars, more comfort and the same handling characteristics.

For race cars with emphasis on handling the damper with remote reservoir is the better option.

The operation of the low pressure twin tube damper remains the same as without gas pressure.

Damping forces are generated by valving in the piston and foot valve and in addition there is the gas pressure exerting a constant force in extension of the damper or strut.

In the normal twin tube dampers, used at the rear of the car the effect of gas pressure is less manifest because the piston rod diameter is smaller in damper designs, but the solution is using a larger diameter piston rod.

A 16 mm piston rod has a cross section of 2 square centimetres so that a gas pressure of 6 bars delivers 12 kg or 26.4 pounds of lift

The piston rod diameter used in this design is most likely 12 mm; the positive lift produced by the gas pressure is by using 6 bar Is 6 kg or 13.2 pounds because the cross section is one square centimetre.

The low pressure twin tube gas damper is used mostly in passenger cars but has gained popularity also in racing applications but for the high performance race application we prefer to use the single tube damper design with external reservoir.

18

Function Of The External Gas And Fluid Reservoir

Single tube dampers have a packaging problem, the fluid and gas reservoir take up too much length to fit in a modern race suspension.

The gas volume is too small for the damper to function properly ...

Aside from the packaging problem the single tube damper without external reservoir functions completely different from the single tube damper with external reservoir.

In the single tube damper without external reservoir, fluid and gas volume are placed length wise (axial), therefore the size of the piston rod diameter (displacement) is a limiting design factor.

Especially for race applications, the external reservoir is a better solution for packaging problems.

A damper design without an external reservoir appeals to most design engineers, there is the perception that locating the reservoir somewhere in the vicinity of the damper is problematic

Jan Zuijdijk

THE EXTERNAL RESERVOIR IS MULTI FUNCTIONAL.

An external reservoir is not just an expansion vessel absorbing the fluid volume displaced by the piston rod.

The external reservoir is a device performing several different functions in addition to being a fluid and gas container.

Gas and fluid volume are separated by a floating divider piston; in this design the top housing of the reservoir contains the adjustable low and high velocity valving, operated by adjuster knobs of different diameters.

Fluid displaced by the piston rod is passing through the adjustable valving system located inside the reservoir first before flowing into the reservoir space, this fact changes the fundamental operation of the gas damper with external reservoir completely.

Whereas the mono-tube gas damper relies completely on the presence of a high pressure gas volume in order to function properly, the gas damper with external reservoir does not need high gas pressure to function but can be varied between 5 bar 70 lbs. to 25 bar, 350 lbs offering a large range of lift provided by the gas pressure.

The fluid reservoir space is expandable by moving the floating piston as much as necessary by the fluid volume entering the reservoir into the gas volume.

The gas volume is compressed only as needed to absorb the fluid volume and is large enough to accept the amount of fluid without a perceptible increase in pressure.

There is always a small increase in pressure but it is so minimal that the driver will not notice the difference, most of the suspension travel in a modern race car centres around a 5 centimetre or 2 inch stroke, alternating below and above the ride height line.

Using an external reservoir offers the possibility to relocate valving into the valve housing of the external reservoir

The valve stack on top of the piston which in a mono tube design damper generates the compression damping force is no longer necessary and can be replaced by a single one way valve with a light preload, this valve is now only functioning as a one way or non-return valve while the fluid displaced by the piston rod is transported to the valving in the reservoir.

The fluid meets resistance through the valving system, comprised of small orifices and a spring loaded blow-off valve where the compression damping is now generated.

In fact in the compression stroke, the damper now works as a twin tube damper system, the valving in the external reservoir acts as an adjustable foot valve in which both low and high velocity damping are adjustable.

It is very important to recognize this change in operation, the compression damping forces in this design do not depend on high or low gas pressure, the damping function is completely independent from the gas pressure function, having a distinct and positive effect on handling.

It is important to recognize the difference in damping technic; the damping forces generated in this system will give a different feel to the car but also improved grip, traction, power down, acceleration braking and cornering qualities if adjustments and gas pressure are used correctly.

The generation of damping forces and function of the gas pressure providing lift are now completely separated from each other explained in the following chapters.

What are the advantages of this system?

In this system, gas pressure is not needed to generate compression damping forces.

The gas pressure can be varied to provide variable lift to help carry the weight of the car.

The gas pressure is now an important adjustable set up tool.

Compression damping is generated by displacement of the piston rod volume from the cylinder into the external reservoir, so the compression forces are generated by fluid passing through an adjustable valving system inside the reservoir.

The diameter of the piston rod is not critical for the operation of the damper function but the diameter can be selected to offer the largest fluid displacement.

The rebound damping is still generated by the large displacement of the piston diameter.

IN THIS SYSTEM GAS PRESSURE IS NOT A FACTOR WHICH IS NEEDED TO GENERATE COMPRESSION DAMPING FORCES.

We have to look at this system as a combination of two damping systems in one, in which the advantages of both systems are combined, and the disadvantages are removed.

It is a combination of a mono-tube gas damper with an outside fluid / gas reservoir connected, so that the mono-tube damper body is completely filled with damper fluid, and the compression valving is placed outside the damper body.

Since the gas pressure is not needed for the hydraulic damping function, it can be used for other purposes like providing a variable lift, depending on the gas pressure used. (Remember a mono-tube gas damper needs 25 bar, 340 psi to function) the gas hydraulic damper with external reservoir can vary the gas pressure between 5 bar to 30 bar which is a fundamental difference.

The lift is generated by the gas pressure working against the piston rod area, at JRZ; I have pioneered the use of a large diameter piston rod, the theory behind it being that the suspension travel of a modern race car is restricted to small movements under and over the static ride height line.

Smaller suspension movements need a larger fluid displacement to generate responsive damping, while at the same time the large piston rod area provides lift.

Some people "test "the damper, by trying to compress the damper by hand and think that the fact that they cannot push the damper in by hand, means that the damping is too heavy to function offering a harsh ride, well the problem is not in the gas pressure or in the damper, but in the perception of the person who thinks he can push more than 200 pounds.

Most people know or find out that they cannot push down on a coil spring but still want to "test "the damper in which the gas pressure exerts a force of 80 kg or 176 pounds.

25 bar of pressure working against the piston rod area of 3,8 square centimetre produces a lifting force of 95 Kg or 209 pounds.

It is impossible for a human being to compress a damper by hand, people are not strong enough to do that, and it would be the same as compressing a non-gas damper with a coil spring mounted on the damper, which is not possible either.

Don't think that using this kind of lift it is not necessary, many times in recent major race events, like the 24 hour race of Daytona and many more other long distance races have we used the addional maximum lift to tune the suspension of a race car providing better traction, handling, improve tire wear and win races that way.

As soon as the damper is mounted in the car and the sprung weight of the car is resting on the spring damper unit, the lift produced by the gas pressure is part of the suspension and carries part of the weight of the car.

The damping function is separated from the lift and that is exactly why the dyno has to be zeroed before testing a gas pressurized damper so that only the damping performance is visualized on the graph without interference of the gas pressure.

The lift produced by the gas pressure helps the spring carry the sprung mass or weight of the car while controlling and dampen the mass of the unsprang weight at the same time.

Helping the spring carry the sprung mass of the car is obvious and logic, when a spring needs to carry 600 pounds of weight

The spring needs also to carry the dynamic weight which is the sprung weight plus the force with which the car is pushed down by the aerodynamic devices, (down force of a car in motion),

The dynamic weight can be quite higher than the static weight; it is obvious that the spring rate needs to be calculated in line with the higher load and natural frequency of the dynamic weight and damping values have to respond to the dynamic forces of the car in motion.

PISTON AREA RESPONSIBLE FOR REBOUND DAMPING FORCES.

The large diameter piston (in most designs 44mm) 1,73 inch is responsible for the generation of rebound forces using a respectable 11,4 square centimetre working area displacing fluid through the system.

The working area of a piston in rebound is the total area of the piston minus the area of the piston rod, in this case a 22 mm piston rod diameter, so the total piston area of the 44 mm piston is 15.20 CM square minus 3.8 cm square equals 11.4 cm square.

This is the working area with which the damper operates in the rebound stroke.

When the damper is extended by the sprung mass of the car in motion, fluid which is trapped between the piston and the guide is pressurized till the pressure is strong enough to open the blow-off valve located underneath the piston.

The large displacement area above the piston is an advantage of the mono-tube design damper over the twin tube damper.

The mono-tube damper with external reservoir works in extension (rebound) as a single tube damper with a large displacement area, and in compression as a twin tube damper with the advantage that the compression damping forces are generated in the valving system located in the external reservoir.

There is a tendency in the race industry to minimize the size of dampers to in some cases pencil like units, car designers like to leave as little space for a spring damper unit as possible in order to keep the car profile as low as possible.

Downsizing of one of the most important suspension components, contradicts the purpose and the expected performance of the damper.

What a damper needs to perform well is fluid displacement.

VEHICLE DYNAMICS AND DAMPING

Fluid displacement can be achieved in different ways, one can use a small damper with very small piston rod and very long suspension travels

The only problem here is that those long suspension travels are not available in modern race car suspensions but are a thing of the past.

In most race applications, certainly open wheel cars, the suspension travel is restricted to 2 a 3 centimetre or a little over 1 inch with a fairly high frequency, except when running over large bumps like curbs etc.

A damper with a small body and small piston rod can never generate the compression damping forces necessary to respond to the high load input.

As a result of downsizing, the extreme small dampers cannot coop with the damping requirements of the particular car in motion and people come up with add on devices to counteract the lack of damping.

A normal size damper with ample displacement capacity to deal with the dynamic suspension requirements of a race car, special open wheel cars, does not need additional help from other devices to cancel out vibrations which should not be there in the first place

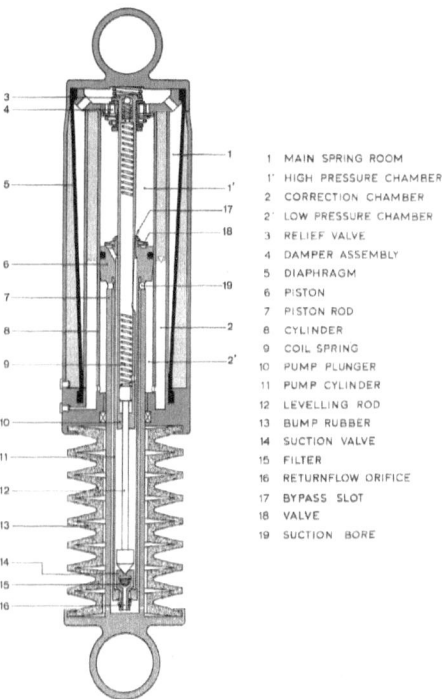

1   MAIN SPRING ROOM
1' HIGH PRESSURE CHAMBER
2   CORRECTION CHAMBER
2' LOW PRESSURE CHAMBER
3   RELIEF VALVE
4   DAMPER ASSEMBLY
5   DIAPHRAGM
6   PISTON
7   PISTON ROD
8   CYLINDER
9   COIL SPRING
10 PUMP PLUNGER
11 PUMP CYLINDER
12 LEVELLING ROD
13 BUMP RUBBER
14 SUCTION VALVE
15 FILTER
16 RETURNFLOW ORIFICE
17 BYPASS SLOT
18 VALVE
19 SUCTION BORE

19

EXPERIENCE WITH COMPRESSION DAMPING

Ccompression damping (bump) damping is more important than the rebound damping for a good handling car.

There is a wide spread opinion in the Automotive and race car industry that rebound damping is the prevailing factor for controlling the dynamic forces of a vehicle in motion.

In my early career and for a long time thereafter, attention was always focused on designing rebound damping forces to control the movement of the sprung weight of a vehicle.

Whether it would be a race car, sports car or a road car, we always designed the rebound damping characteristic to keep the sprung mass controlled under the most diverse conditions.

Even today, the overall perception in engineering is that the compression damping should be as light as possible.

The sprung mass of a sports car reaches a natural frequency of approximately 1,5-2 hertz depending on weight, sprung mass and spring rates of course.

The compression or bump side was never considered to be much of a factor in the complete picture of suspension control; it was believed that bump damping control had to be kept at a minimum in order to achieve better ride qualities.

Just enough compression force visible on the graph but not more.

Looking at the effects of damping in a different way I came to the conclusion that the weight of all suspension components, the un-sprung weight, was a disturbing factor which was not taken into consideration by most engineers.

I have told the story of how I arrived at that conclusion and have designed the JRZ series of dampers following that philosophy with great success.

The un-sprung mass when un-dampened is a greater disturbing factor to the ride and handling of a race or sports car or any moving vehicle for that matter than the sprung mass, a fact that is not yet widely accepted but true.

Consider the mass and the way it is agitated by road undulations moving un-dampened in the frequency of the tire, feeding high forces at a very high frequency into the suspension of the car, whether it is a sports, race car or road car.

From earlier suspension developments I experienced the benefits of using gas pressure in combination with hydraulic damping to control the un-sprung mass.

The gas pressure works against the cross section of the 22 mm piston rod and delivers a lift to carry the sprung mass without a noticeable spring rate.

Gas pressure is active over the total length of the suspension travel, keeping the tire contact patch in firm contact with the road surface.

The gas pressure is variable and can be adapted to track conditions and is a valuable tuning tool.

Development work in ride and handling with major sports car manufacturers and race cars ranging from showroom stock, sports cars, Trans Am, prototype cars, Formula One through Indy cars and all sorts of open wheel cars etc. have provided Me with a wealth of experience accumulated in the JRZ damper design of today.

Even recently testing this principle on drag cars proved the concept to be right, the car gained better traction with each increase in gas pressure and increased compression damping, contrary what those people had been used to.

It is my experience that a race or sports car using struts or dampers with a large displacement diameter piston rod and variable gas pressure gain traction or grip dramatically when the low velocity compression damping is used to tune the handling of a car.

For a good understanding, when I talk about a large piston rod, I mean to use two aspects of the large piston rod and that are fluid displacement and the gas pressure lift.

The low velocity compression damping and gas pressure controls the un-sprung weight in a very effective way, eliminating most of all the wheel chatter and jerky movements of the suspension felt in a car.

In addition, the gas pressure has a direct positive effect on handling and controlling the un-sprung mass of a car as well.

Most engineers think of the rebound damping first as the main tool to change the handling of a car.

Experience obtained in testing sports and race cars have thought me that the rebound force is not as important as a well-tuned compression damping / gas pressure force.

The common approach to vehicle dynamics by the majority of Engineers and designers is directed at the wrong part of suspension and damping, the extension (rebound) movement of suspension and damping, controlling the sprung mass.

In the early days of damping design, we have been thought that the sprung mass was the most important component to be controlled by the hydraulic dampers we were using in those days.

My experience is that controlling the un-sprung mass of a car contributes much more and in a much noticeable way to a perfect handling and smooth riding car.

I used the method of questioning everything that we had been doing in the past and what was accepted as common science by the majority of automotive engineers.

Why keep doing it this way if there is little or no progress, and why keep doing the same thing over and over again when the results are always the same and less than satisfactory?

Examples:

When working as a young engineer, it was, in addition of rendering technical service to the racing team, my job to work with sports car manufacturers, tuning the damping and suspension to the specifications and satisfaction of the factory ride engineers for the next model year sports cars.

Our most common assignments were customers as Ferrari, Lamborghini, Alfa, Porsche etc.

After a couple of years working under the direction of a senior engineer, I was send on my own assignments.

It meaned that now I could use my own insights and experience, using less conventional methods without the controlling factor of widely accepted industry standards.

Using different methods meaned that any design change had to be made in accordance with the factory design philosophy and production capabilities.

All factory test drivers, whether at Porsche, Ferrari, Lamborghini or Alfa were very skilled and dedicated to the important job of creating the best performing sports car, when the tests were finished and the results approved, the car would go into production for the next model year and no major changes were allowed after that.

Damper characteristics were recorded; the results laid down in a protocol for the quality control department and from there on the supplier had to adhere to the specifications.

It occurred to me that we always did the same thing, namely using and modifying the rebound damping characteristic to arrive at a good handling and comfortable riding car.

Although I realized that we did not use all the options we had available, it took courage to deviate from the proven engineering philosophy practised over decades and on which the company design philosophy (and success) was based.

JAN ZUIJDIJK

Coming back home from each test session, we had to report to the owners directors who were also practical engineers going over the results and how they were achieved.

I always have had the greatest respect for the two brothers Arie and Kor De Koning, founders of the Koni damper company, Porsche in Stuttgart was one of the favored customers of Koni in those days.

Sometimes, mr Kor made time in his busy schedule to be present at the ride development test and spend several days at the Nurburgring with Me to personnally see what progress was made, He did not hesitate to take a lap around the Ring in the race car with Herbert Linge, Porsche factory test driver to experience first-hand how the car handled.

Companies have quitte rigid rules for damper and damping design which is not so strange if one considers that damper design changes have to be in line with the production methodes and design philosophy of the car company.

My first chance to use a different approach came when I was doing ride development at the Porsche factory in Stuttgart-Zuffenhaussen

For every change we made to the valving of the dampers, spring rate or antirollbar rate, the car had to be driven over the open road and over a certain parcours knowen and laid out by the test driver/engineers

The test track at Weissach did exist but was not ready for high speed testing.

Our test driver used to be mr Herbert Linge, a very experienced test driver who also participated in long distance races like the 1000. KM of the Nurburg ring, le mans 24 hour and the Targa Florio.

All kinds of situations were simulated when driving the car at high speed over the autobahn.

Sudden lane changes, running the clover leafs of the autobahn crossings at high speed, hard braking, hard accelleration, it was tested in hair raising manouvers.

One important tests was an accelleration test stopping the car at the bottom of a hill on a mountain road, put the car in first gear, accellerating uphill at full throttle.

VEHICLE DYNAMICS AND DAMPING

The road surface was full of holes and bumps, so the rear suspension, the driven wheels, shuddered and moved noisely, violently, up and down under the torque of the engine, while the tires could not coope with the power, winding up and slipping alternately, when they reached the maximum traction level so that the car gained speed, fish tailing with big swoopes up the hill, it was nearly undriveable and dangerous in accelleration under those conditions.

Modifying the rebound characteristic did not make any difference but made it only worse.

Back at the workshop of the Porsche factory, i was expected to have a solution at hand for this problem because the suspension would not be approved for production.

Thinking about how to solve this problem, it occurred to me that we were always working on the rebound side of the damping curve, modifying the characteristic in the low and high velocity damping range adding orifices, reducing pistonbores because we believed and were thought that the sprung mass would compress the springs, storing energy of motion in the springs when being compressed.

On the return stroke the spring will give back the energy by extending itself above ride height, the rebound damping resists this extension movement converting energy of motion into heat or thermal energy, controlling the excess movements and velocity of the sprung mass.

Whatever changes we made, there was a difference in handling but it did not solve the powerdown problem.

In the mean time we let the unsprung mass free to move and accellerate, generating a lot of kinetic energy with little or no damping control, and if there is very little damping there is very little conversion of kinetic energy which means that the undampened energy can move the suspension assembly at a much higher velocity and travel than is necessary or desired becoming a very disturbing factor.

I decided that in order to solve this problem it was time to use a different approach ignoring the vested phylosophy and start working with the compression side.

I modified the compression damping to a much higher level.

In those days, the possibilities of modifying the compression damping were restricted, we had a choice to take a footvalve out of a selection of 10 numbered and preset footvalves for the twintube dampers we were using, 0 being the lowest damper setting and 10 being the highest, the damper still had a 12 mm piston rod with a small displacement volume so the suspension needed stroke and velocity to be effective.

Whereas all engineers before Me always used the lowest available footvalve, I broke with that tradition and decided to increase the compression damping dramatically using the number 4 footvalve which was a break with past engineering practices.

The next test with the modified compression characteristic was a revelation, the car, in the same uphill test was all of a sudden more docile, had a lot of traction and grip and was much better to control while the ride quality had not suffered but was even better than before.

Following this test we went to the autobahn clover leaf and ran around the curves at maximum speed, the car was handling beautiful with even better ride qualities than before.

It showed that much of the discomfort experienced in a moving car originated directly from the un-dampened or insufficiently dampened kinetic energy generated by the un-sprung mass.

The positive results inspired us to continue working in this new found direction and the end result was a nice riding and handling car.

Taking this idea a step further and thinking about the theory behind it, I decided to use a damper with a larger piston rod next time, a larger size piston rod could displace twice as much more fluid over the same stroke length than the damper we had used before.

Using a larger piston rod meaned also using a larger size damper which was the next step up in piston size.

This was easier said than done because the consequence was that the cost of the heavier damper was also higher and had to be accepted by the financial department but at Porsche quality always prevailed over cost.

The piston rod diameter in this design was 16 millimetre (0,63 inch) and the piston size was 33 millimetre(1,3 inch).

The larger piston rod size displaced twice the amount of fluid per centimetre stroke than the damper we used before in compression, so the compression damping was double effective compared to the damper we used before.

My objective was to test the larger displacement of the piston rod in compression and see whether my theory worked.

The next tests were a lot better again, special the uphill power down test greatly improved to every bodies surprise and satisfaction of the factory engineers.

This convinced me that there was more to gain in the compression damping/un-sprung mass than was ever expected before and we should rethink the theory of the rebound forces being the main controlling factor for handling.

The tendency before was trying to go lower and lower in compression damping arriving at almost zero damping in compression.

MONTEVERDI SPORTS CAR.

The next occasion to prove my theory in practice came when one of our customers needed help with the ride of his car.

Peter Monteverdi was à sports car builder in Switserland building a limited number of high performance sports cars using our spring damper units and was very unhappy with the ride and handling of the car.

At the time I was doing development work at Porsche in Stuttgart when I received a call from our sales office with the request to stop by at Monteverdi to see what the problem was, Mr Monteverdi had complained about the ride and handling of his cars.

We took the service truck to Switserland, arriving late in the afternoon but Mr Monteverdi was eager to show me what was wrong with the car so I was put in the car right away

After a couple of minutes in the car driving through the city streets we noticed that the car had a very poor ride quality over bumps and cobble stone streets jumping over every little bump.

The ride was very harsh and uncomfortable, while the tires were leaving the road surface frequently with any small bump or dip in the road.

We asked them to take the dampers of the car and opened the dampers in our well equipped service van.

Checking the damping performance on the dyno, I noticed that the damper had almost no compression damping; the force line ran almost over the zero line of the graph but had a lot of low speed rebound damping.

I decided to increase the compression damping dramatically on the front as well as rear suspension, reducing the low velocity rebound damping at the same time.

After we made the modification the dampers were checked on the dyno again and reinstalled in the car.

It was already late in the evening but Mr Monteverdi could not wait and we went out of the door for the next test and rode the car for 30 minutes through the same streets as before.

The result was astounding, Mr Monteverdi was so surprised that his car could be so comfortable and well handling at the same time that He could not stop driving the car and talking about it, He send a telex (the faximile had not been invented yet) to the factory right away saying that He finally had the damper valving He needed for the car and that He finally was very satisfied with the performance of the suspension and the ride of the car.

This incident proved again that my theory was correct and to go forward in the same direction with other development projects.

Up to this point, the damper tuning was following the theory that the sprung mass was the most important component to be controlled.

The first time we turned away from this approach, the ride and handling improved dramatically

FORMULA ONE

We also rendered service to the Formula One series and long distance racing like the 24 hours of Le Mans, the 12 hours of Reims, the 12 hours of the Nurburgring Etc.

The problem with this kind of suspension engineering is that you have to rely on the information you receive from the drivers and in most cases through intermediation from the team engineer which also meaned that the information from the driver could be modified by the engineer and could be different, adding His own vision to the handling and balance of the car.

I vividly remember one occurrence at the track of Watkins Glenn in the USA; I was in charge of the Koni service team USA.

I worked with the team of Ferrari, the chief engineer had not arrived yet and the engineer replacing him was very cooperative following my advice about the setup.

Maybe the fact that he was less experienced in suspension tuning was an advantage.

The cars were notorious under-steering and the drivers complained about this as one of the main factors for the slow lap times.

The car would not enter the corner well and the driver needed too much time before getting on the throttle accelerating out of the corner.

I did what recent experience had thought me reducing the rebound settings dramatically at the front while increasing the compression forces at the same time without changing spring rates (we did not use gas as a setup tool yet)

The under steer disappeared, the car went much faster and lap times improved to the point that the drivers were happy with a competitive car.

Then the chief engineer arrived and undid everything we had worked so hard for to arrive at a well-balanced and handling car.

He changed everything back to his personal specs.

The results were dramatic, when the race started, the cars were passed by the competition and finally one car came into the pits, the mechanics took the front dampers of the car and showing them to the fans on top of the pit boxes, yelling "ammortizzatore" claiming that the damper were the source of bad performance, which in a way was true, but they failed to say that not the damper but the totally wrong damping characteristics and wrong adjustments dictated by the chief engineer was the source of the problem.

Needless to say that, when an F 1 car stops in the pits for a suspension change, any chance of a good finish are lost.

We were still using dampers without gas pressure so the reservoir held air at atmospheric pressure in the expansion space.

The design philosophy of Koni for which I worked at that time did not allow any use of gas in the reservoir.

The directors, owners of the company found that using gas in their design was an admission that the twin tube design was inferior to the gas damper design so we were not allowed to use it.

There was a trend in the racing industry to favour mono tube gas dampers over hydraulic dampers in those days, because the gas dampers could be mounted upside down, offering the advantage that the un-sprung weight would be lower, which is debatable because the steel piston rod assembly is usually heavier than the aluminium damper body.

Working with the Ferrari F1 team in Europe, one of the drivers asked me to design a damper which could be used upside down for his formula 2 car.

I constructed a hydraulic damper with an extra cylinder inside so that the foot valve would draw fluid through the extra cylinder.

The air in the system had to be removed carefully but it worked real well on the dyno, and when the dampers were installed, the driver liked it so much that he never gave them back.

When we met again later at one of the Formula One races, He told me that the car handled much better and was quicker through the turns winning races, He liked them very much.

The driver was Lorenzo Bandini who drove formula two races in between His Formula One commitments.

Sometimes it seems that part of the improvement is mentally but the stopwatch does not lie.

Later in my career when living and working in the USA we were forced to start using gas in the twin tube damper for road car applications by the competition.

The so called soft gas or low pressure gas damper was being promoted as the new developmentment.

Competitors were applying gas pressurized twin tube dampers to road car manufacturers.

At the time I was cooperating with a new development project at Ford SVO, developing the suspension for the SVO Mustang.

It was to be a limited edition of a new design Mustang with an intercooled turbo Engine.

Because the competition used gas in a twin tube damper, the so called low pressure gas damper, we had no choice but to develop our own version.

Since I had my own development department in the USA, it was not too difficult to design a low pressure gas strut and damper without interference of our parent R and D department.

The car was equipped with external adjustable Mc Pherson struts in front and an independent rear axle with twin tube external adjustable dampers in the rear.

The struts and dampers were adjustable in rebound only as was normal in those days.

The gas pressure was 6 bars, which was why the dampers were called low pressure gas, contrary to the single tube gas damper using the 25 bar pressure necessary to generate high compression forces.

The 6 bar gas pressure however a positive effect on the ride and handling of the car had and was easier to tune than ever before.

Throughout all the testing we did, it showed that the ride quality of the twin tube damper with the help of gas pressure was at a much higher level than the plain twin tube damper used before.

Especially the front struts worked very well, the Mc Pherson struts had to be designed with larger piston rods, because the struts have to take the side loads induced into the front suspension when cornering, break torque under braking and acceleration when installed in a front wheel drive car.

The piston rod was 22 mm in diameter.

The force exerted by the gas pressure onto the piston rod controlled the un-sprung mass of a car much better than with non-gas struts, non-gas twin tube struts and dampers use the atmospheric pressure to operate so they easily activate.

Cavitation means that in the extension stroke (rebound stroke) the piston rod volume is extracted faster from the working cylinder than the damper fluid can follow, creating a low pressure area underneath the piston.

The low pressure area will be quickly filled with air so that there is a mixture of damper fluid and air inside the damper, there is always a certain percentage of air in the damper fluid, which in a low pressure area comes out of the fluid, accumulating as a gas or air bubble.

It is therefore very important that the flow through the non-return valve of the foot valve is not restricted in anyway, there should be sufficient bores and the preload on the spring should not be too strong in order to fill the cylinder on the rebound stroke.

20

Development of the Self-Leveling Suspension for the Porsche 911 and Ferrari 365 2+2

The experience obtained with development work on a project with so called self-levelling spring damper units gave me an indication of the potential of using gas pressure in combination with hydraulic damping.

The self-levelling components we developed were designed to take the weight difference of a sports car between completely empty and fully laden and designed to keep the car level within different load variations.

This means that the car totally empty or totally laden with full tank of fuel, maximum luggage and full load of people including the driver was to stay at the same ride height level.

In addition the vertical g force of the impact of a big bump had to be absorbed without bottoming the suspension out and damage the internals of the unit.

The rear suspension of the Ferrari 365 2 + 2 was designed with one damper in front of the rear axle and the hydro pneumatic self-levelling unit at the rear of the rear axle on each side.

The spring in the rear suspension was a low spring rate coil spring and was designed to carry the basic weight of the car while the self-levelling suspension unit provided the lifting force to compensate for the added weight of fuel, driver, people, luggage and other objects.

The unit increased the lift by pumping fluid from a secundairy chamber to the main chamber, increasing the gas pressure inside as a result of the reduced gas volume and in the process increasing the spring rate and keeping the car at ride level.

The pumping action came from the suspension movements as soon as the car started rolling over road undulations the pump action started.

The unit had three main positive characteristics or qualities.

The hydro pneumatic self-levelling unit could carry a minimum load of 150 kg and pump up to a load of 550 kg per unit.

When the car was at ride height, the pump action stopped because the plunger of the pump had a tapered flat part ground in the side of the plunger.

At ride height this tapered part would rise above the edge of the pump cylinder and draw no longer oil from the reservoir but keeping the main chamber pressurized so that the car remained at ride height

The pressure in the main chamber would remain at the same value until weight was taken out of the car, the levelling valve was lifted of its seat and pressure was released reducing the lift.

The high pressure was kept in the main chamber by a tapered valve closing off a large orifice at the bottom of the cylinder.

Only if the car body would rise above ride height would the tapered valve be pulled of the valve seat and release some of the pressure flowing back to the low pressure chamber.

The valve would close immediately after some pressure was released because the car body would respond by lowering itself and close the valve.

The load capacity depend on the size of the piston rod and gas pressure inside the spring unit.

The second important property was keeping the natural frequency of the suspension at the same level unders all load conditions.

By selecting the correct gas volumes, the natural frequency of the sprung mass could be kept at the same level in laden and un-laden position, the natural frequency adapted to the load variation automatically

When the load increased, the self-levelling unit would respond pumping more fluid from the low pressure chamber into the main chamber increasing the spring rate at the same time, so that the natural frequency was kept at the same level under all conditions.

When the load is increased and the spring rates kept the same, the natural frequency is lower, but a lower natural frequency results in longer suspension travel with a chance that the suspension will bottom out on bad roads.

In a conventional suspension, spring rates have to be higher than necessary in order to carry passengers, luggage and a full tank of fuel.

In the self-levelling suspension units the natural frequency can be lower because the ride height is kept at the same level.

The frequency remained the same in laden and un-laden condition, keeping suspension travel the same in laden and un-laden conditions.

This condition was responsible for the very comfortable ride of the car whether the car was empty or fully laden.

Weight did not change comfort and handling because the suspension compensated for the load difference adjusting the spring rate to the different load condition keeping the natural frequency at the same level.

The third quality, I think was not very well understood at that time but most important.

When a normally (steel springs) suspended vehicle runs over a big bump the energy accumulated in the springs will propel the car body (sprung weight) far over the ride height line while the damper is working to slow down that motion.

A steel spring delivers the same force back after being compressed plus that the dynamic weight of the car changes direction on the rebound stroke and the dynamic weight of a car in rebound is less than in the compression stroke.

The self-levelling suspension unit did exactly the opposite, when the car equipped with the self-levelling units ran over a big bump and the suspension would extend itself above the ride height line, the unit released a little of the excessive pressure reducing lift and spring rate and the car would settle back immediately at ride height without the need of an hydraulic damper.

It did not have the big excursions of a conventional suspension but the ride was very smooth but firm.

The difference is the way energy of motion is converted; a conventional suspension will give back all stored energy and needs to be dampened by hydraulic damping, generating heat.

The self-levelling unit, by blowing off the excess pressure out of the main gas chamber and therefore reducing extra lift, did not need the heavy hydraulic damping, because the energy of motion was stored as gas/hydraulic pressure and could be dissipated without disturbing the ride of the car, the self-levelling suspension performed much better than a conventional suspension but the absence of damping forces controlling the spring energy manifested itself in the exceptional smooth ride.

Another observation was the absence of damper tuning, when there is no energy to be converted, there is much less need for hydraulic damping.

The self-levelling had a certain amount of damping build in but it was basically a restriction of the return flow during the blow-off of pressure.

The self-levelling suspension was developed for the Ferrari 365 two plus two, a larger than usual car from the Ferrari designers.

The self-levelling suspension unit was located behind the rear axle and would take care of the added weight like fuel, luggage, passengers etc.

All production cars of this type have been equipped with the combination damper/spring unit and the self-levelling unit.

The lower natural frequency manifested itself also with better traction and excellent grip.

I drove the car throughout the tests many times and the car displayed a very nice ride combined with excellent handling.

The total absence of tire chatter of the driven wheels under acceleration was also a remarkable sensation.

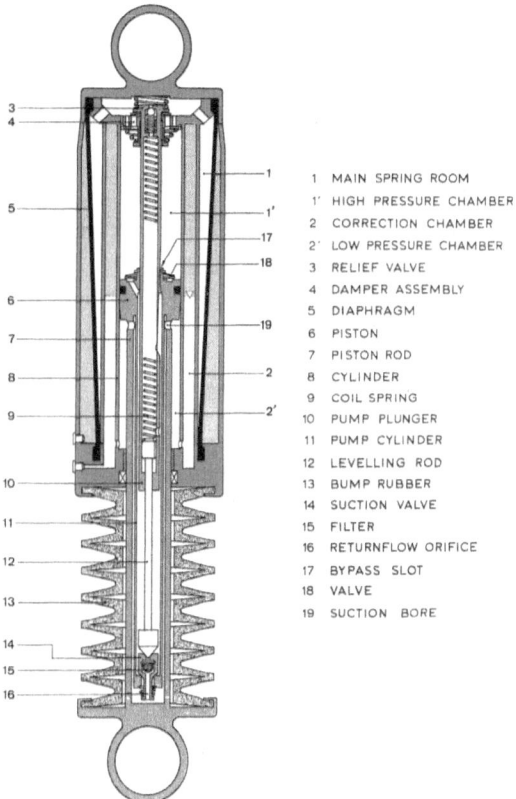

1. MAIN SPRING ROOM
1'. HIGH PRESSURE CHAMBER
2. CORRECTION CHAMBER
2'. LOW PRESSURE CHAMBER
3. RELIEF VALVE
4. DAMPER ASSEMBLY
5. DIAPHRAGM
6. PISTON
7. PISTON ROD
8. CYLINDER
9. COIL SPRING
10. PUMP PLUNGER
11. PUMP CYLINDER
12. LEVELLING ROD
13. BUMP RUBBER
14. SUCTION VALVE
15. FILTER
16. RETURNFLOW ORIFICE
17. BYPASS SLOT
18. VALVE
19. SUCTION BORE

The same design self-levelling unit was also developed for the Porsche 911, although the application was reversed, in this case the unit was built into the Mc Pherson strut because the Porsche had a rear engine the most weight difference between laden and un-laden condition manifest itself at the front of the car, where the fuel tank and luggage space were located.

In this application, the unit did not have any help of a coil spring but it was the only suspension component the car had.

The unit worked very well, the car had a very nice ride and due to the natural frequency being kept at the same level under all load conditions the ride was very comfortable but firm.

The best property I observed was the fact that the front of the car settled immediately after running over a bump without the usual up and down motion.

When running over a bump in the road or a rise, the front would lift and settle immediately without any wild after moves.

The fact that the kinetic energy stored in the suspension levelling system was released rather than the energy stored in a coil spring pushing the front up and being dampened by a hydraulic damper made all the difference in the world.

In fact using suspension travel to produce lift and carry part of the weight of the car was an early form of a Kinetic Energy Recovery System (KERS) and far ahead of its time.

Unfortunately, only a small number of self-levelling suspension units made it into the production cars at Porsche, they did not rely on one supplier and since the second supplier could not come up with an acceptable product, the project was cancelled but I never forget the lesson learned from this development.

This development was a big step ahead in suspension design and I have never understood why this design did not receive a better reception by the car manufacturers.

When I developed my own damper design, I have incorporated the positive aspects of gas pressure into the combined damper/gas pressure damping system.

I used the gas pressure to provide lift to the suspension, using a 22 millimeter piston rod to provide substantial fluid displacement and the gas pressure to provide lift varying from 6 bar 22,75 kg to 25 bar, 94,75 kg.

Choosing the gas volume large enough, the gas pressure in the damper design does not increase the spring rate.

The lessons learned working on the self-levelling suspension were not lost, I always kept in mind the good properties of this design and always looked for ways how to use those properties in a positive way.

One aspect of the experience was the lift produced by the gas pressure and the amazing positive effect it has on car handling.

VEHICLE DYNAMICS AND DAMPING

When developing my own design damper, I have made use of that one aspect combining hydraulic damping with the effects of lift produced by the gas pressure in combination with a large diameter piston rod on one occasion; I could prove my theory making the difference between failure and success.

Cross-Section of Self-Leveling Hydropneumatic KONI Suspension Strut

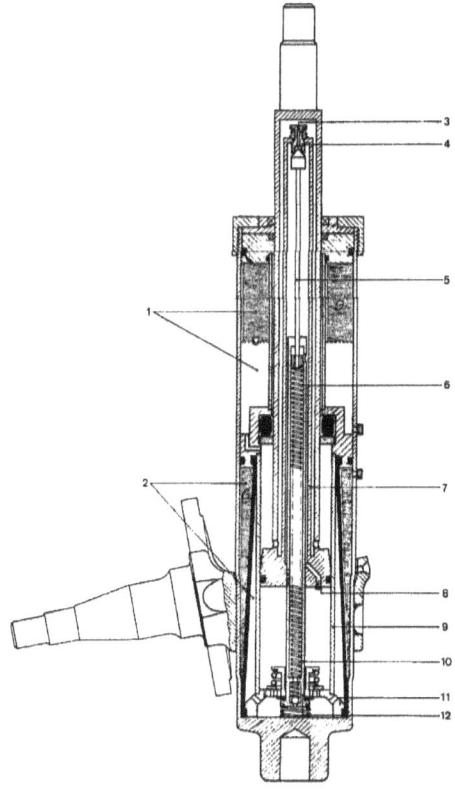

1. Low pressure chamber
2. High pressure chamber
3. Return flow orifice
4. Suction valve
5. Needle valve
6. Spring
7. Pump cylinder
8. Pressure valve
9. Main cylinder
10. Pump rod
11. Damper housing
12. Overload valve

G = Gas cushions

21

Front Wheel Driven Race Cars

Vehicle Dynamics of Front Wheel Drive Race Cars

Front wheel drive race cars require a different approach.

Working with the Alfa DTM car, there was a problem with tire wear on the front axle

For example the car could run 10 laps at Silverstone and blow one of the the front tires, in particular the tire which was taking the most inside corners.

As I analysed the situation, the car rolled excessively in the front, lifting the inside tire almost of the pavement, the reason being a combination of too high rebound forces low spring rates and low antiroll bar settings at the front of the car.

When the car made a left turn the left front wheel would get light but the differential would keep driving the wheel scraping the inside edge of the tire over the pavement.

That was the reason that the tire wear occurred at the inside edge of the tire and wear through within 10 laps, when the tire was worn through the tire would blow.

This had nothing to do with tire compound or quality, the tire was built well and could run a whole race without problems, it was simply a problem of the suspension setup and geometry.

When a team wants to test your product it is common practice to base-line the car and driver with the components they are using up to the moment that the driver declares the car ready to make a change.

After the car has been adjusted and base-lined which takes mostly the morning session they will switch to the new components, it this case my design damper.

Again, my theory is to support the outside wheel in a turn instead trying to lift the inside wheel with a strong antiroll bar setting and heavy rebound damping forces.

In my opinion, the lift produced by the gas pressure combined with higher bump or compression forces keeps the outside tire better in contact with the road surface, controlling roll the right way. When the dampers were installed on the car and the driver went out for the first laps, we made some adjustments to the damper settings and leaving everything else the same, the driver went out for the 15 lap test.

The lap times were excellent and when the car came in way after passing the critical 10 lap limit, the tire which used to wear excessively and blow every 10 laps was examined by the tire people and team engineer.

The wear pattern showing the inner edge of the tire totally worn before was not there anymore as a matter of fact the tire showed hardly any wear looking like new.

Again the test proved my theory and design to work well planting the tire contact patch over the complete width on the road eliminating all the problems previously encountered by scraping the inner edge of the tire in every corner.

FRONTWHEEL DRIVE RACE CAR.

I was contracted to work on a suspension development program with Nissan Europe, the car was a front wheel drive developed in England and raced in Spain, the car was driven by Luiz Sala and Eric van der Poele, two outstanding drivers competing in major international races all over the world.

The car had not been fast enough to win the championship and through the engineering staff, I was contacted to see whether we could get the car to work better (make the car handle better and faster)

The engineering department was in England so I went to England for a meeting to get the measurements and particulars about the damper units.

We made a set of dampers to the specifications and measurements and went to a race track Albacete Spain for the test, test driver was Luiz Sala, former formula one driver, the test track is short but has all ingredients to tests all aspects of car handling.

The engineering staff attending the test was from England in addition to the regular spanish team members.

The complaint was that the front and rear end had a life of their own, in other words, the drivers explained that the rear end did not seem to be attached to the front wheel driven car meaning that the handling of the front axle was noticeable but the rear-end just followed along

Not responding to any changes made to adjustments.

We spend half a day setting up the car to the best results they had achieved over the past season and when the car was base lined, we started working with the set of dampers I had prepared for the test.

This set was the basic gas hydraulic design, double adjustable with external reservoir, large 22 millimeter piston rod and 45 mm diameter piston.

When the car went out on the track for the first time, the driver did not report much difference so we started to build up the spring package first which was necessary because the suspension was on the soft side, increasing the compression damping step by step the handling started to improve.

Again the key to the improvement in handling was the combination increased low velocity compression damping and increasing gas pressure, especially the rear end came to life with the improved damper settings and spring combination.

A front wheel drive car is more difficult to tune because the rear-end of the car is just following the lead of the front.

The natural response of race engineers is to make the rear suspension as light as possible

With as result that the car has a tendency to over-steer, the rear-end will break out at every corner.

When I was confronted with this phenomenon, it was a little difficult to increase damping and use more gas pressure in the damper to keep the un-sprung mass controlled better.

It feels like going against better knowledge but in reality, the handling of the car improved with every increase of compression damping and gas pressure, resulting in.

Lower natural frequency

Better traction

More grip.

Less tire wear.

We have proven that as recently as January 4 2013 in Daytona where the car needs a lot of support to withstand the vertical g forces of the banking and down force produced by the wings.

The natural frequency of 4 hertz requires a spring rate of 1300 pounds/inch because due to the g forces in the banking the dynamics weight of the car is much higher.

22

Development Of The Dashboard Adjustable Damper

As we have seen before, designing a correct damping characteristic is not some kind of black art, for people unfamiliar with damper design and tuning it might look that way but dampers respond to laws of physics like any other mechanical device, present in the suspension of a race car.

The damper is designed to counteract, dampen or slow down the suspension movements in a controlled way, so that the suspension is not abruptly stopped over a short part of the travel, but the damping should be applied over a long enough part of the travel, so that the suspension returns to the correct ride height in a controlled way without going over or under the ride height level. As a matter of fact, it is applying the correct dose of energy conversion over the shortest but least disturbing suspension travel.

The damper controls excessive compression of the spring during the compression stroke and the extension of the spring on the rebound stroke so that ride height is restored in a controlled way.

Controlling the compression stroke is very important, when the damper dampens the compression stroke well, the suspension travel will be reduced almost by 50 % which became evident when I tested with an Indy car at Mid-Ohio.

When the original dampers were replaced by JRZ dampers with external reservoirs, the car used only half the suspension travel as before, with better traction, handling, power-down and cornering.

VEHICLE DYNAMICS AND DAMPING

Being able to run a car lower means making better use of aero dynamics and a faster better handling car.

To make a damper work in harmony with all other suspension components It is recommended that the designer has a thorough understanding of hydraulics and flow of fluid in combination with experience of working on and driving high performance cars on the road and on the track.

It is this combined experience that makes the engineer an expert in the world of vehicle dynamics, damping and suspension tuning.

Having driving experience makes it a lot easier to relate better to driver comments about car behaviour and handling.

Designing a characteristic making an optimal contribution to the ride and handling of a car, one needs a thorough understanding of In fact what happens is that when the suspension is compressed by the sprung mass of the car running over road bumps, the energy of motion or kinetic energy is absorbed and stored in the compressed spring for as long as the mass is on its compression stroke or downward travel, storing more energy than needed to carry the sprung weight of the car.

The reason for the excess energy accumulated in the spring is that the suspension is compressed by the sprung mass of the car in addition to the vertical acceleration (up and down movement) the resulting force is more than needed to carry the weight (mass)of the car.

When the mass of the car changes direction and moves in extension again, the compressed spring reacts, extending it self, it has absorbed more energy than necessary to carry the weight of the car plus the extra dynamic weight, overshooting ride height if the suspension would not be dampened.

The rebound damping force slows down that extension travel converting the surplus of energy of motion into thermal energy or heat.

The extending damper pulls the piston through a cylinder filled with fluid, the flow of fluid through a spring loaded valve creates a hydraulic pressure drop on the discharge area of the piston, and this pressure drop from high pressure to low pressure generates the rebound damping force which In turn generates heat.

The art of designing a correct damping characteristic is to apply exactly the correct dose of damping force at any velocity to release the energy stored in the spring in a controlled way over a certain travel of the suspension available.

This is easier said than done; the load input of a vehicle in motion is not constant but is variable depending on road conditions, down force, cornering, speed etc.

For those who like to look at force velocity graphs and have a clear prediction of how the line should increase with the increasing velocity I have a disappointment.

A damper working in the suspension of a moving vehicle does not follow the nice force line drawn as recorded on a damper graph from the dyno, the damper moves intermittent under and over the static ride height line in small increments with occasional peak loads when a car hits a bigger bump or runs over a curve, under braking, cornering forces etc.

The amount of damping used is sometimes expressed in a certain percentage of critical damping.

Critical damping is the situation at which the damping force is so strong that all the kinetic energy is absorbed in one movement and the damper prevents any further movement of the suspension.

A 30% of critical means that the damping rate applied is 30% of what the force would be to stop any movement of the suspension in any direction; this does not indicate in what velocity range the 30 % is applied.

This is all theoretical, it is much easier using common sense to arrive at a correct damping characteristic because the sprung mass is not activated at the same rate and frequency but it depends on the variable road condition and changing mass of the car as we said before.

It is nonsense to calculate a percentage of critical damping for a race car or even for a road car because the load input depend on road or surface conditions and are so variable that only a wild guess can be made, just using a more powerful engine can change the requirements completely.

Critical damping is not something the average racer has to bother with; it diverts the attention away from the real issues at hand so let's not make things more complicated than they are.

The correct damping characteristic can be determined approximately and has to be tuned on the road or race track.

Different drivers have different driving styles with a direct influence on how the characteristic should look and feel on in the suspension of the car.

Many times, I had a perfect tuned suspension with a lot of traction and grip when the driver wanted a more crispy handling which is in fact giving up a little on grip and traction making the car respond more direct to steering, braking and power input so that the driver can use throttle and brakes in cornering more selective or to his liking.

The answer to this condition is changing to higher spring rates all around combined with higher rebound settings, special with heavier race cars this can be helpful for drivers who have been used to a stiffer setup and crisp handling.

I have encountered this phenomenon more than once, the dampers, antiroll bar have been tuned to the correct settings but the spring package remains the same, in some cases the team engineer does not want to make changes to the spring rates He is used to, but if the damper setup provides optimal grip and traction, the spring rate has to be in line with the higher grip level otherwise the car will respond slower to directional changes and cornering.

The initial roll is slower and the car will settle into the steady state cornering later into the corner.

The spring rates however have to be in harmony with the increased grip level in order for the car to respond quicker to driver input.

A crisper handling car usually means that the driver has to work harder but if that is the driving style one has been used to, it might be quicker but also more tiring.

For long distance races, it might be more prudent to keep the more docile setup so that the driver is less tired, the tires don't wear so quickly and the car performs better over a long period of time with the same set of tires.

A good way of testing the effects of damping is to take an external adjustable damper, and apply a lot of low velocity rebound damping to a normal road car.

The effect will be that the car when riding over a bump, will throw one out of the seat, because the car body will stop abruptly its upward movement, while the vertical acceleration will propel your body upward with so much force that your body will leave the seat cushion.

His is because the rebound damping slows down the motion of the suspension to abrupt but the human body when not held down by a seatbelt continues the upward motion and lifts of the seat.

It is the vertical acceleration of the car and the human body which is felt as very uncomfortable.

The same is true for a race car; the driver will find the same situation very uncomfortable,

The tires will lose contact with the road, have poor traction and overall bad handling

This situation does not contribute anything to a better handling; it is only uncomfortable for the driver and wears out the car quicker.

We developed a damping system which was adjustable from the dashboard, the damping forces were really adjustable in rebound and in compression.

The system had 7 positions and each position closed off the free flow of fluid from the top of the piston to the lower part of the cylinder.

The damping was adjusted by means of hydraulic fluid running through a small diameter hydraulic tubing connected to the front struts and rear dampers, running through the hollow piston rod moving a little piston closing or opening orifices.

The hydraulic force, adjusting the dampers was distributed to all four wheels and worked on the rebound side and simultaneously on the bump side to a lesser extend.

The results were astounding, when the dampers were adjusted at maximum the ride was hard like a brick but handled well on the race track.

I brought the car to the Michigan speedway where I was testing with Patrick Racing, the Indy car team I worked for at that time with Emerson Fittipaldi and Bruno Giaccomalli driving the Indy cars.

Bruno Giacomalli and I drove the car with the adjustable dampers on the Michigan speedway and although it was a high performance sports car with turbo engine we could drive full throttle all around the track without any problem or adverse handling problems.

The surprise came when the adjuster was turned down to the minimum position, it felt like the car started floated on air when the strong rebound and compression forces were released and the suspension able to move and working again.

I had the car for a couple of weeks to develop the suspension and tune the damping, I made frequent test drives over the highway and on race tracks changing the damping characteristic in order to find the optimal suspension performance for the car.

This was actually the perfect way to adjust the suspension and if we would have made the front and rear adjustable independently it would be an even a better tool to tune the suspension.

The problem is that under race conditions, the driver wearing driving gloves does not have the fine feel in his fingers to make delicate adjustments with a little knob.

When we tried it on a race car the driver came back into the pits holding the adjuster knob in his hands, so that idea was not very practical besides the fact, it is not a good idea to leave damper adjustment to the driver who needs to give his full attention to driving the car and stay on the track.

23

EFFECTS OF DAMPING ON CORNERING

Damper settings can effect cornering of a race car to a great extend.

Keeping in mind that a race car wants to go around a corner at the fastest possible speed without getting of the line, keeping good traction, grip, acceleration and pull the maximum of G forces without breaking out front (under-steer) or rear (over-steer).

Damping forces effecting the optimal cornering of the car must be found in a combination of the compression or bump side of the damping characteristic and a certain amount of rebound damping.

Compression damping force in combination with the lifting force delivered by gas pressure Increases the lateral grip to a great extent, balancing the handling of a race car front to rear is nothing other than applying a combination of low speed /high speed damping and reducing or increasing the gas pressure at the front or rear end of the car in addition to spring rates and roll bar settings

The idea is to support the outside wheel/tire in a corner instead of an antiroll bar lifting the inside wheel, which is getting light already because of the weight transfer from the inside to the outside wheel.

First, check the rebound settings of the dampers and when they are to high let's say above 50% of the total range, position, the front or rear of the car will have a tendency to drift out of the corner so if the rebound settings are close to the maximum adjustment range reduce rebound to lower settings, somewhere halfway the adjusting range.

VEHICLE DYNAMICS AND DAMPING

Only when you adjust the rebound to lower settings than required is it possible for a driver to feel the effect of the lower velocity rebound forces, and can be adjusted back up to find the optimal setting.

Check the spring rates used on the car, if very high spring rates are used, adjusting damping forces will have almost no effect on the handling of a car because the springs are overriding inputs from other components responsible for the handling.

Lower the spring rates in balance with the natural frequency of the sprung weight.

Spring rates should be in line with the natural frequency of the sprung mass because the spring rate is a variable component of the formula calculating the natural frequency.

The natural frequency can be calculated using the formula:

1 divided by 2 pi x square root out of spring rate divided by the weight of the corner of the car: spring rate =hertz

If the natural frequency is way high, it might be considered to reduce spring rates to a more acceptable level.

Also the spring rate split front to rear should be in balance with the weight distribution front to rear and of course the velocity ratio.

The spring package is the foundation of the car and needs to be addressed first before making all kinds of adjustment to other components because it is easy to get lost in the many adjustments possible.

When spring rates and rebound damper settings have been addressed it is time to focus on handling in cornering.

Leaving the rebound adjustment at a fairly low level, it is best to use a couple of laps to base line the car and see where everything is set at . . .

First address the driver's main problem.

When the car under-steers, apply more low-speed damping, low speed damping is working two ways in the damping characteristic.

One click of low speed damping increases the nose angle of the damping, graph while also increasing the high speed damping force at the same time

This is inherent with the laws of hydraulics, increasing the low speed damping means closing of one orifice so there is more fluid passing through the blow-off valving system, this results in more damping over the high velocity range of the damper.

The change of high speed damping forces is significant; it can amount to a 100 % damping increase in high velocity damping at the same time that the low velocity damping is increased by closing down orifices.

The driver should notice the change when going out on the track again, if the driver reports an improvement, try another increase of one click compression damping.

In order to keep the car balanced front to rear, make the same adjustment at the rear.

If the low speed damping is close to the maximum of the range, increase the high speed damping one click at the time front and rear.

One has to be very methodical making those adjustments, it is so easy to make adjustments that do not make any difference in handling but when not corrected, they might interfere in a negative way with the setup.

Always make one adjustment at a time, if you make more than one adjustment at one time, it is difficult to find out whether the change made was an improvement or not and what part of the adjustment made the difference.

First how does the car react when entering a corner, under braking?

If the rear of the car is unstable under braking going into a corner, most likely the rear low velocity rebound damping is to strong, or the flow through the piston is restricted, it does not allow the suspension to extend itself quick enough to keep the tires in firm contact with the road surface, causing the rear tires to lose the ability of transmitting braking power to the road surface.

The car will be very unstable and over steering when accelerating out of the corner

The car will be loose and unstable under braking as well as acceleration.

The best answer to this problem is to reduce rebound damping, if the high and low velocity damping can be adjusted separate, reduce the low velocity first and bring it back to the halfway point.

If the piston bores are restricting the flow of fluid (bores are too small in diameter), the piston should be replaced with one with larger bores and perhaps a less heavy valving stack or if the shim stack is preloaded, reduce the preload.

The suspension will be allowed to extend itself at a much faster rate, keeping the tire contact patch in contact with the road surface under braking.

The first reaction to a handling problem is increasing the rebound damping side of the damping graph, believing that this is the way to get a faster and better handling car.

The truth is that the rebound damping does not bring the improvement they are looking for at all.

Too much rebound damping affects the handling in a negative way; rebound damping is designed to dampen the motion of the sprung mass of a car, whatever car it is.

An empty road car can take four people weighing approx. 70 kg each

A full load of fuel, 60 liter weighs 50 kg, luggage 100kg, this brings the total added weight to 430 kg.

Recalculating the natural frequency changes the mass with 430 kg which means that the natural frequency will be lower, because the spring rate stays the same in the equation.

The direct effect of a lower natural frequency is that the suspension travel increases, because all other components of the equation remain the same, except the mass increases with 430 kg.

The spring rate of the springs carrying the sprung mass are one component in the equation calculating the natural frequency of the sprung mass of the car.

The natural frequency of a sports car is approximately 1,2 hrtz

The natural frequency of the sprung mass of a race car is 3 to 5 hrtz depending on what kind of race car it is, cars running a banked oval have to calculate also the vertical g force caused by running at high speed on the banked part of the track.

The static weight (mass) of a car can vary considerably from the dynamic weight or mass.

The static weight is when the car is sitting still either laden or un-laden.

The dynamic weight is the static weight + the added weight as a result of down force when the car is traveling at speed and of course the vertical acceleration of the sprung and un-sprung mass in motion.

This is the challenge when setting up a race car for a super speedway.

The spring rate will be much too high for the weight of the car when parked in the pits but at speed, the added weight or down force will change the equation and bring the natural frequency down to a lower level.

The advantage of tuning for the speedway is that the speed of a car is almost constant, except for moving around other cars.

The disadvantage is that the handling of a car running in the wake of another car on the speedway can be abruptly upset by the sudden absence of down force.

Cars tuned with maximum mechanical grip have the advantage in this situation, race cars which have been tuned with emphasis on down force are at a disadvantage.

For road courses, the situation is more complex, because cars running a road course have to negotiate high speed corners, braking, low speed corners, while the benefits of down force in slow corners declines or is absent.

Braking heavily for corners, accelerating away from corners, for a car racing on a road course, there is a different force input introduced from outside sources than a speedway car.

It is our experience that sufficient damping should be applied at the piston velocity range of 0 to 5 cm/second

VEHICLE DYNAMICS AND DAMPING

The low velocity damping controls the energy so well that it prevents the suspension from reaching higher velocities so that the higher end of the curve becomes less important, it is not necessary to read the damping forces at the higher velocities because the suspension will not reach those piston velocity levels.

Most drivers we work with report that the gas/hydraulic system with large piston rod diameter responds to every single click of adjustment.

This is a result of a build-in strong increase in compression damping, at a very low piston velocity.

Designing the damping system, I have looked at every part to be optimal functional; the damping system is capable of damping the highest load inputs, without giving a harsh feel to the ride of the car.

The basic damping system performs so well that every time when testing with a new team and the original dampers are replaced with the JRZ suspension system, the handling and laptimes improve instantly.

During damper and suspension tuning it is advisable to keep the antiroll bar at a soft setting or use an antiroll bar with a low spring rate.

The antiroll bar is a device responding to displacement only, this in contrast with dampers which are responding to displacement and velocity.

On a race track with a rounded surface the antiroll bar effects the handling in a negative way because it connects the right and left suspension, forces induced in one wheel will affect the other side, causing the car to swerve from left to right and is difficult to keep in a straight line.

The suspension supported by the lift of the gas pressure works the opposite way.

The lifting force in the left and right suspension works independent from each other so that the tires have a better road contact with the road surface without interference from left to right.

Testing at the Sebring track made that evident, running the old part of the track with rounded surface, the car we were testing with was directional unstable at high speed.

The heavy antiroll bar made the car swerve from lef to right and the only way we could get it out was to use a smaller antiroll bar, increase the gas pressure and compression settings.

The gas pressure and the heavier compression settings supported the wheels independently without interference from side to side.

24

Forward Traction

The engine in a race car with lots of horse power available, transmits the power through the transmission, differential, suspension, damper and finally through the tire contact patch onto the pavement or road surface.

A race tire is a very impressive large round piece of rubber but the contact patch transmitting power to the track surface is only a small strip of rubber

Despite the most advanced electronics to get the last ounce of horse power out of an engine, it is still that small piece of rubber that has to be in contact with the road at all times in order to transmit engine power to the pavement

Any time a tire leaves the road surface even for a split second, power is not transmitted it is that simple.

A race car wants to accelerate and slowdown in the shortest possible time because a quick lap time Is all important for race car drivers and teams.

When a tire is not in firm contact with the road surface, no matter how powerful the engine maybe, the car will not accelerate quickly or go around the turns in a controlled way.

Bad or intermittend surface contact causes wheel spin, lack of grip, little lateral grip, slow acceleration bad braking performance.

When the tire lifts off the road surface, even for the shortest possible time, the engine power cannot be used to propel the car forward at full power or slowing the car down.

The JRZ designed gas hydraulic damper plays a very important role in this complexity of power inputs, keeping the tire in firm contact with the road surface.

There are several conditions preventing optimal tire / road contact, using a high spring rate will make the suspension more rigid, causing the tire to skip over the road surface with small amplitudes but high frequency.

This pattern causes a tire to heat up quickly, every time the tire leaves the road surface for the slightest time period the tire tread will wear, and the time the tire spend up in the air there is no transmission of power possible, the tire will lose its peak performance very quickly.

Putting the power down is especially difficult when a car is accelerating uphill in low gear on a very uneven road surface like street circuits.

The old street circuit in downtown Detroit was a prime example of how important power-down and acceleration really is.

The track made a sharp right turn behind the pits, the Indy cars had to accelerate uphill full throttle in lower gear for a short period of time, it is this part of a race track where the difference in handling is manifested and where we could make the difference, when Emerson Fittipaldi, driving the Patrick Racing Indycar was using our design gas hydraulic dampers for the first time.

On the starting lap Emerson got stuck between the inside wall and another car, He limped back to the pits for a new wheel and tire, resumed the race, was hit for a second time and spun around, re-entered the race and with all the time lost, charged to the front to take the lead in the next to last lap, finishing in first place.

Emerson continued to win three races in a row, Detroit, Portland and Cleveland which was convincing evidence that the new dampers we had designed made a major contribution to the handling of the car.

We had tested with the team in Florida the week before on a small race track, Jim mc Gee, the Patrice Racing team manager immediately saw the advantage of the new damper setup, Jim was taking corner times and

VEHICLE DYNAMICS AND DAMPING

looking at the greatly improved corner times came to the conclusion that Patrick racing had a race winning setup and a lot more grip and traction than before.

Jim immediately called Pat Patrick to report the improvement; Pat made a deal for the rest of the season, keeping the damper setup exclusive during the 1989 season.

Jim used to call me when I was the director of engineering at Koni America every time He needed bump rubbers, He could not use our dampers at that time because that would have been a conflict of interest but when He took the position as Team manager at Patrick Racing and I started my own business, We could work together on damper development for Indy cars without restrictions.

Jim has a fine sense for new developments, improvements that can make the difference in racing; Patrick Racing won the championship that year with most wins, most pole positions, most laps lead and most mileage raced.

The Toronto race was one more occasion at which our damper design and race engineering experience made a big impression and the difference between winning and losing.

The race engineer came up with a different plan for this race, took our dampers of and had a set from the competition on the car when I arrived at the race track, this incident shows that you can never lean back and relax but always have to be on top of things.

It was a big disappointment, the car was 3 seconds slower and the driver very unhappy about the handling of the car, the car was not competitive at all but we worked out the practice session

During lunch break, Emerson confronted me with this situation in the race truck, He asked why the handling of the car had changed so much for the worse, he was not aware what happened and when I explained the situation he asked me what to do.

I said the solution is right in front of you because the set of our dampers which I had prepared for this race was on the work bench inside the truck and I estimated that there would be at least 2 seconds to be gained in the suspension setup.

Emerson insisted that the dampers were installed and went into the last practice before qualifying, I was wrong with the 2 seconds, the lap-time improved with 3 seconds and Emerson was on the pole again.

Emerson was always a gentlemen and very appreciative for good work, He stopped the car right in front of Me after his qualifying run for the pole and stuck two thumbs up in the air giving credit where credit was due I guess.

The high torque of the engine in low gear causes a high torque on the rear tires and suspension.

The car will squat and load up the tire, the tire bites into the pavement, when the tire has reached its maximum load capacity, the tire will slip / spin for an instant and unload the tension which has been build up in the tire, the suspension reacts with a violent vertical acceleration when there is to much damping In rebound and little in compression or bump, this process repeats itself at a fairly high frequency till the car has built up enough speed to shift in a higher gear,

The effects of this condition can be greatly reduced using the correct compression damping settings at low piston velocity in combination with lift provided by the gas pressure in the reservoir working against the piston rod cross section, (controlling the un-sprung mass through a combination of gas pressure and compression damping)

Sufficient compression damping at low velocity needs to be dialled in using the low velocity damping adjuster knob.

The best way is to turn the rebound damping back to position 10 Increase the low velocity damping to position 4 and the high velocity damping to position 8

Every time the low velocity compression damping is increased, the high velocity compression damping increases simultaneous.

This is because when more orifices are closed of (which is what happens when the low velocity compression damping is increased) a greater volume of fluid is passing through the valve stack which shows up on the graph as higher damping force at the high velocity range, increasing the nose angle(steeper)

In fact what happens, the increased compression damping, combined with the higher gas pressure, dampens or slows down the motion of the un-sprung mass, which is felt as better control, better handling and a better planted tire.

Dampening the un-sprung mass is the most miss understood and most under estimated but the most important part of handling.

Think of the mass of wheel, tire, brake rotor, brake calliper, control arms and other components, how much kinetic energy is generated when brought in motion and not properly dampened.

If the un-sprung mass is not dampened correctly, the suspension will use more travel in compression than necessary.

With the rebound force adjusted to a firm setting, the damper will not release the compressed spring quickly enough so that the suspension is still below ride height when running over the next bump and will compress even further.

The chassis is jacked down into the suspension by the overpowering rebound forces in low velocity and is cause for a bad handling car.

When this condition is present, dynamically the car rides below static ride height, because of the jacking effect, the rebound force will not let the suspension extend itself quick enough so that the ideal ride height never will be reached.

As a result the car runs lower, but also the spring is more compressed than is necessary to support the weight of the car, the ride will be harsher than needed, having a negative effect on handling and tire wear.

It is under those conditions that the well-tuned spring/ damper combination manifest itself in letting the car accelerate in a very controlled way with good grip.

The performance of a race tire depends on several factors, rubber compound, air pressure, carcass design; tread with, tire wall design etc.

For technical information and characteristics, it is best to contact the tire the engineers, present at most race events.

They can give information about pressures, camber and tire characteristics

The tire is the only component in contact with the road surface and therefore it must be our objective to create the best possible condition for the tire to do its job.

When a suspension is set up correctly, it greatly affects tire wear in a positive way.

On one occasion, we were racing in Atlanta, when our car came in for a pit stop and tire change.

After the car had left the pits, the tire engineer asked the mechanic, could you show me the tires which just came of the car?

The mechanic showed him the tires but the tire engineer said, no, I mean the tires which just came of the car on the last pit stop, well, the mechanic said, they are the tires which just came of the car.

After a full race stint, the tires looked like they had not even been used on a race car, this shows the benefit of a well-balanced, handling car, and the car was leading the race for the whole duration and finished in first place.

25

Damping and lateral grip

The most important part on a race car is the tire, the tire contact patch is the only part of the car in contact with the track surface and is subjected to the input of many different forces, torturing the tire to an extend that it is sometimes hard to believe that it can stand up to all the abuse which it is subjected to.

To make optimal use of the lateral grip performance of the tire all other suspension components need to be tuned to peak performance and working together in harmony, the tire needs to be planted onto the pavement in order to withstand the G forces generated by the centrifugal force of the vehicle at high speed ovals and weight transfer to the outside of corners.

For the front axle it means that the tire has to transmit braking power, roll around the corner, and transmit G forces without slipping or sliding sideways causing under steer or over-steer.

For the rear axle it means that the tire can transmit brake power, take the lateral g forces in turns transmitting engine power and all that on demand in split seconds.

Lateral grip depends on the input of several conditions.

In general it is the roll stiffness of the front or rear end of the car which causing the car to slide around a corner.

Cornering performance is determined by optimal tuning of the combination of spring rate, antiroll bar, rebound damping, compression damping and tire pressure, special the compression damping is very important for the cornering of the car.

The tire is an elastic device, the tire contact patch is subject to constant change, inputs from driver, steering, braking, acceleration, lateral G forces, and vertical G forces all change the contact patch of the tire in motion, one look at the tire in action by an on-board camera is enough to respect the tire designers for ever.

The more mechanical grip the tire can deliver onto the pavement desto more severe the tire loads are.

Aerodynamic forces (down force) high speeds ad more load to the tire.

Many times I have experienced that a driver reported under steer in slow corners, which is a mechanical grip problem, the solution to this problem was in almost all cases, increasing the low speed compression damping forces in front.

Increasing the low speed compression damping consist of closing of a small orifice, offering more support to the outside tire.

Increasing the low speed compression damping with one click and closing of one more orifice means reducing a very small percentage of the available flow is sufficient to stop the under-steer and make the car turning into the corner well.

Better grip of the front tires reduces the slip angle of the tires causing the tires to follow the direction they are pointed in rather than sliding sideways, which is under-steer, so that less tire wear occurs.

Also camber, toe in or toe out have an effect on the performance.

Sometimes, camber settings are used to mask a real handling problem, increased camber is used to make the car turn-in better into the corner.

Sometimes you see cars running with 4 degrees of negative camber, the front tires running on the inner edge because some people believe in excessive camber settings, this kind of solution bears the risk that the tire wears excessive on the inside edge and slows the car down in a straight line speed.

When suspension and damping are tuned well, those excessive camber settings are not necessary, after all tires are meaned to run on the full tread and not just on the inside edge and a stand up tire rolls better with less roll resistance.

Excessive camber settings cause rolling resistance and can slow down the race car at high speed.

The way the temperatures are distributed about the width of the tire gives valuable information about over-steer, under-steer, camber settings, wheel spin, traction grip etc.

The tire factory engineers provide valuable information to the driver and engineer when they take the tire temperature each time the car comes into the pits, they can read the wear and evaluate the temperature distribution over the width of the tire.

The question is, how does one respond to this information into a set up change improving the performance of the car.

The low velocity rebound damping in combination with the resistance to roll of the anti-roll bar can be very helpful stabilizing the race car in cornering at speed, the faster the car travels around the corner desto faster the lap times will be.

However the combination of high antiroll bar settings and heavy low velocity rebound damping is counterproductive to get a car faster around the corner.

The added force of antiroll bar and rebound damping results in a very roll stiff car producing under-steer at the front, and over-steer at the rear of the car.

Those combined settings make the inside tire lift of the road surface when cornering, because the high rebound forces do not allow the suspension to extend itself quick enough after encountering a bump so that the tire lifts temporary of the road surface at small intervals.

I have looked at this phenomenon from a different point of view.

I have found a better way of controlling the cornering of a race car, asking the question why are we doing the same thing over and over again, if the results are not any better today than they were yesterday.

The combined weight of the inside wheel plus suspension is not enough to stop the car from rolling, it will only result in lifting the inside wheel of the road surface.

The answer to the question is to support the outside wheel, instead of trying to lift the inside wheel, it works a lot better, it keeps the inside wheel as well as the outside wheel much better in contact with the road surface, so that both tires perform optimal in cornering, allowing the car to travel much faster around a corner without sliding.

I once designed an anti-roll bar system, controlling the roll of a car by supporting the outside wheel in the corner, having a lateral sensor switch on an additional antiroll bar per outside wheel, while the inside wheel was left free to move and run over curbs undisturbed.

The suspension of the inside wheel could move freely so that no forces were transmitted from the inside wheel to the outside wheel.

This was a great improvement, the cornering speed (time) was much faster without the inside wheel upsetting the lateral grip by transmitting (disturbing) the suspension movement from the inside wheel to the outside wheel.

Running in a straight line both additional antiroll bars were not activated so the car could run over bumps without the left and right wheels interfering with each other.

The car with this system performed perfectly but was not allowed in the race; it was deemed to be an electronic controlled suspension and outlawed by the race organisation.

The experience obtained with this design however showed the way to take a different approach in controlling the cornering forces of a moving car.

In addition to the anti-roll function of the system, antidive and anti-squat could be programmed in using the signal of a longitudal G sensor.

I am sure that there are huge benefits to be found in such a system and one day it will be used but the racing community needs to see the benefits and change the rules.

The antiroll bar is a torsion bar connecting the left-hand side wheel with the right-hand side wheel, the bar resists suspension travel from the left

VEHICLE DYNAMICS AND DAMPING

suspension to the right hand suspension, normally when the car runs in a straight line the bar does not do anything, except that with a heavy roll bar setting, movements of one side will affect the other side, when one wheel runs over a bump in the road, the force will be transmitted to the other side by the bar and affects the straight line stability.

The function of the antiroll bar is to prevent excessive roll of the chassis in cornering by resisting deflection of the bar but it is clear that there has to be roll and deflection before the antiroll bar can react.

Excessive roll is a matter of opinion, the antiroll bar is a passive device, it is a torsion bar and responds to displacement (roll) so if the chassis does not roll an antiroll bar is not necessary.

Roll depends much on the combination of spring/rate, damping force especially the low velocity rebound damping.

A better way of preventing roll is supporting the outside wheel in the corner instead of trying to lift the inside wheel of the pavement by the antiroll bar.

I have pioneered that development, my experience with gas hydraulic suspension units showed the way to a better solution.

One of the problems developing better suspension systems is that the smaller companies do not have the resources to spend the time and money for a development program while large damper companies are more focused on production numbers.

Much of the better ideas are not worked out because of this, while professional racing teams are not interested until the suspension or damper companies present a ready product, and they probably would like to try it if as long it is free of charge.

This is not a formulae for success, the ideal situation is when a damper R&D department can work together with a larger organizations like I was privileged to do when I did development work with Ford, Roush racing and GM chassis development.

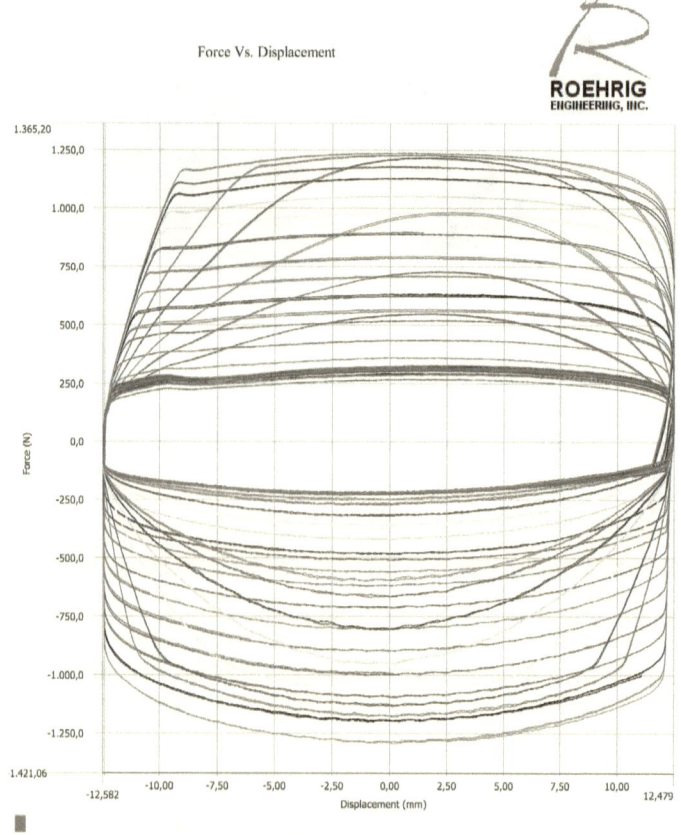

Compression and rebound adjustments

26

Compression damping of the four way adjustable damper

Compression damping control and adjustment of the four way adjustable dampers.

During the compression stroke, damping forces are generated by displacing fluid out of the working cylinder, passing through the compression valving into the remote reservoir.

The fluid is forced out of the cylinder by the piston rod volume entering the cylinder based on the principle that no two solid or fluid bodies can take up the same space at the same time

The fluid is forced through a high pressure hose, passing through the compression valving system into the remote reservoir containing fluid and gas volume separated by a divider piston.

Separation of gas and fluid is necessary to prevent gas and fluid to mix, causing the damper piston to travel through a mixture of gas and fluid, resulting in unpredictable damping forces and handling.

The compression valving system operates in two stages.

At low velocity, from zero to 2 inch per second, the fluid flows through pre measured orifices into the reservoir.

With increasing velocity the flow becomes too large for the orifices to handle, the pressure increases, forcing the blow-off valve to open at a pre-set preload, this preload is variable by means of the adjuster knob, exerting more or less spring force onto the blow-off valve by turning the knob clockwise or counter clock wise.

Turning the knob counter clockwise, the damping force will increase, turning the knob clockwise, the damping force will decrease.

This is simply because the adjuster is right hand threaded and pulls the springs closer to the valve increasing the spring load.

The amount of fluid displaced is equal to the piston rod volume, which is 3,79 Cubic centimetre per cm stroke by a piston rod diameter of 22 mm, for comparison reasons a 14 mm diameter piston-rod displaces 2 cubic centimetre per cm stroke.

The new four ways adjustable damper is equipped with a large 22 mm diameter piston rod, standard in the JRZ product range for large fluid displacement at low velocity.

This is very important for suspension design in modern race cars, the suspension of a modern race car operates at small amplitudes but higher frequencies, so for the damper to be effective in those small suspension movements it is necessary to displace as much fluid as possible per mm stroke.

Some other brand dampers are using smaller diameter piston rods, depending upon the gas volume they have available, but are in most cases restricted in volume design by the available package length and other measurements, unless a remote reservoir is used.

A damper is a passive piece of equipment, like brakes work only when pressure is applied through the brake pedal.

Dampers work only effectively when the piston is moved through fluid caused by motion of the suspension.

The amount of fluid displaced and the damper valving are responsible for the damping force and characteristic of the damper.

The amount of damping generated in a suspension depends on several inputs.

VEHICLE DYNAMICS AND DAMPING

1. The velocity with which the damper piston is moved through the cylinder

2. The amount of fluid displaced by the piston (Rebound) or piston rod (compression)

3. The velocity of the damper (the damper stroke in relation to the suspension stroke) actually the position of the damper relative to the pivot point of the suspension control arm called the velocity ratio.

4. The damping characteristic builds into the damper which is a result of the valving system.

5. The gas pressure acting on the piston-rod cross section, as a helper spring but without adding spring rate to the coil spring.

The high pressure hose can withstand heavy peak loads at maximum adjustment in very heavy terrain for rally cars.

The swivel-banjo connection leaving the hose free to turn 360 degrees eliminates the need to disconnect the hose to reposition the angle on the damper body or reservoir

A swivel connection prevents the mechanic from accidently letting the pressure of the system.

Using a fixed connection it has happened several times that we had to rebuild the dampers even before they had run on the car when a mechanic loosened the hose to clock the hose in a better position and accidently releasing the fluid under pressure.

The pressure hose allows fluid to travel from the damper body to the remote reservoir passing through the orifices and valve system into the fluid reservoir and back.

Flowing back, the fluid opens a one way valve and flows back into the cylinder.

Before this kind of hose was available, the damper dimensions were restricted by packaging space left by the car designer.

The metal armor prevents flexing of the hose so that there is no loss in damping performance due to the hose expanding under high pressure.

Fluid displaced by the piston rod entering the working cylinder travels from the working cylinder through the valve system located in the remote reservoir.

The compression valving system consists of a series of orifices and a spring loaded blow-off valve or a combination of one large orifice and tapered needle valve.

In the JRZ system, the orifices in the compression valving vary in diameter so that the nose angle of the graph is exactly evenly spaced at nice even increments at every click of adjustment.

When turning the adjuster knob adjusting the damping, the orifices are closed of one by one.

The low velocity adjuster knob is located at the top of the valve housing, each adjustment represents one click.

The valving in the reservoir has been redesigned and operated by two knobs side by side, one for the low velocity wnd one for the high velocity damping.

Every click closes of one orifice, there are 14 low velocity adjustment positions.

Closing of one orifice at the time guarantees a consistent step by step increase in damping force, in the initial stage of the damper stroke, the velocity starts from zero so the amount of fluid displacement is low at very low velocity.

During this stage the blow-off valve will remain closed, while the fluid starts passing through the orifices.

The size of the orifices vary from 0.8 mm in diameter to 0.5 mm in diameter, each diameter orifice offer a certain resistance at low velocity.

This is indicated by what is called the nose angle of a damping graph, it is the increase in damping force relative to the piston velocity.

VEHICLE DYNAMICS AND DAMPING

At a certain point in the piston velocity, the amount of fluid is too much for the orifices to pass through, the resisting force increases and the spring loaded valve opens.

The blow off valve lets a larger amount of fluid pass relative to the piston velocity.

This is the point in the damping graph where the line sharply changes from an increasing angle to a more horizontal line.

High and low velocity damping can be changed by rearranging the number and size of the orifices in combination with the valve seat diameter and spring load, this is designing the damping characteristic.

Without changing the combination orifices and valve, the effect of adjustment is visible on the graph, where the line changes with each adjustment.

When the first orifice is closed off, (first click of the low velocity adjuster) the nose angle of the damper will show a steeper line on the graph with each click of adjustment.

When the second and subsequently all other orifices are closed of one by one, the graph will show a steeper angle with each click until maximum adjustment, the graph will than show a

vertical line from zero to the blow off valve opens there are 5 positions in the low speed velocity adjustment.

This system guarantees a very accurate adjustment range of the damping forces.

In this system every click shows always the same increase in damping meaning the damping forces can be reproduced by simply putting the adjuster back in the desired position, in case of doubt of the damper settings all one has to do is to set the adjusters back at the last position known as correct and start from there or adjust from the zero position.

Any time race engineers are confused about the damper settings they can go back to zero setting the adjusters to the adjustments of the quickest run and damper performance will be restored exactly to that same value again.

Compression damping high velocity.

With higher piston velocity, volume of fluid flow increases with velocity, the blow-off valve opens and releases the fluid into the reservoir, the blow-off valve is spring loaded; the spring force can be increased or decreased by turning the high speed knob at the top of the valve housing.

When checking the high speed compression force on a dynamometer keep the low speed adjuster closed.

This way the damper shows only the high speed damping force by every click of increase.

The system shows very consistent increments in increase in the damping force.

Increasing the spring load results in higher damping forces

Decreasing the spring load results in lower damping forces

There are 14 high speed adjustments, each step representing a certain increase or decrease in damping force, marked by a distinct click system.

The adjustment positions are very precise making it easy to return to earlier damper settings of the same damping level as before.

This is important to know, I have noticed that it can be confusing what the high and low speed damping really means and what it does in the suspension of the race car.

It is understandable, time to set up a race car in practice is very short, there is never enough time to do everything needed to make the car perfect, most race engineers know adjustments, but don't have an indebt knowledge of damper technology or how it affects the handling of a car.

Qualifying is a different story.

In qualifying, the car needs to make a very fast lap time within the first one or two laps, so all the suspension settings have to be set to serve the only purpose to heat up the tires very quickly to peak operating temperature within one lap.

Damper settings are critical for peak performance in that situation.

REBOUND DAMPING OF THE FOUR WAY ADJUSTABLE DAMPER

In a four way damping system the rebound damping force is generated by the valve stack in combination with the orifices in the piston rod.

The flow through the orifices is different, the rebound high speed and low speed damping have to go through two separate hydraulic circuits.

The fluid released by the blow-off valve (high speed) is one circuit; the fluid flowing through the orifices (low speed) is a second circuit.

Orifices are open connections between the cylinder space above the piston and the cylinder space below the piston allowing a free flow between the two parts of the working cylinder.

The total area of the orifices should be calculated as a percentage of the total piston area, if this percentage is to small, the result will be that the lowest adjustment setting still will be to hard and a harsh ride.

Low rebound damping is controlled by the orifices in the system, the orifices are adjustable and can be closed of one by one so the nose angle of the damper changes from very shallow to almost vertical.

When the damper starts its rebound stroke from zero, the fluid which is displaced by the piston flows through small orifices into the lower part of the damper cylinder.

The number and size of open orifices determines the increase in damping force relative to the piston velocity, while the blow off valve takes over when the flow of fluid becomes too Much for the orifices to handle, which is usual when the velocity of the suspension travel reaches velocity higher than 5 cm or 2 inch /second, both systems are adjustable.

The low velocity is adjusted by closing off the orifices one by one, when turning the adjuster to a higher setting, with every click the graph shows an increase in damping force with a steeper nose angle every time another orifice is closed off with very consistent and even increments.

The high velocity damping is adjusted by applying more spring load onto the blow off valve when the adjuster knob is turned to the plus sign.

This is visible on the graph by the increase in damping force, every click of adjustment will show a higher line in damping so that there are no differences from one damper to another.

Rebound damping is necessary to dampen the kinetic energy which is stored in the spring.

When the suspension is compressed as a result of running over a bump on the road surface, the compressed spring wants to turn back to its original position and extend itself back to ride height.

When too much low speed rebound damping is applied the spring will not be able to extend itself quickly enough to ride height because when the suspension is compressed repeatedly and the low speed damping force is to strong, the damper will hold the spring down in a compressed position (jacking effect) the spring will operate at a higher force level than necessary to support the car, a race car will respond by a harsh ride over bumps and not handle well.

The car will be difficult to point in the direction the driver wants the car to go, therefore compression damping force is far more important than the rebound force.

The compression or bump force will plant the tire better onto the pavement; result is better traction and better power down.

The car will also use less bump travel so that the car can run lower to the ground helping the aero dynamics stabilizing the car at high speed.

When the car uses less bump travel, the car can use higher spring rates and run lower.

The rebound force should be set high enough to stabilize the car in cornering so that the driver feels the improvement.

When working on the race car setup most engineers select rebound damping as the first choice to counter under-steer or over-steer.

Assuming that cambers, caster, toe's etc. have been set at known settings there are additional tools to set up a race car suspension success fully and balance the car handling front to rear increase lateral grip and traction.

It is important to keep in mind that there are several tools available for tuning a suspension.

1 Spring rates.

2 Rebound damping

3 Compression damping

4 Gas pressure

5 Antiroll bar settings

6 Aero dynamics.

7 Camber, Caster, Toe settings.

It is very important to look at the settings of all components when adjusting the race car setup to a competitive level.

If one components provides an overriding force, spring rates are too high or rebound damping is too high or one of the components are set at extremes there is no way that the suspension can be tuned with success because when the spring rates are too high, adjustments of the other important components have no effect on the handling.

Sometimes it seems that not sound engineering is leading to intelligent decisions in car setup but looking what the other guy is doing and copy it.

Some time ago, I was engaged in a test weekend at Daytona in preparation for the 24 hour race, I had not been involved in the race setup of the GT cars for a while and I was told that the best setup was a very soft front end, this to get the car around the corner in the quickest possible way because they were using JRZ struts and dampers, looking at the results, it was obvious to Me that several tools to improve handling were overlooked.

We were confronted with a situation in which the front spring rates were lowered to a level not providing much support for the car.

The compression damping was at the lowest level of adjustment both the low velocity and high velocity settings were adjusted to low settings.

The gas pressure (one of the most powerful setup tools for traction) was lowered to a minimum.

Only the rebound damping was at an acceptable level.

Gas pressure in a JRZ damper provides a lot of support in addition to the spring rates because of the large displacement 22 mm piston rod diameter and is very effective so is my experience.

It seems a natural reaction of engineers to lower spring rates, lower compression damping settings and lower gas pressure.

When the cars went out for the first time the track was green and the drivers had to get used to the car and track so the first session did not tell us much about changes to be made to the setup

The weather on the first day of testing was cold, windy and occasional light rain.

VEHICLE DYNAMICS AND DAMPING

When the track conditions improved the cars went faster and the drivers complained about under-steer conditions in slow corners and lack of lateral grip in the front end.

The engineers reacted in trying to even lower in the settings, while it was suggested that the spring rates should be lowered once again.

Because it is not my style to bud in and make random changes without having a clear idea why, we evaluated the situation and came to the conclusion that it was time to once again use the experience we gathered from the many test sessions with all kinds of race cars and said to the engineer.

We lower the rebound settings to position 6 increase the gas pressure to 18 bar, increase the low velocity compression damping to position 4 and the high velocity compression damping to position 8 at the same time, because these are changes which can be made quickly and evaluated in a time frame of three laps.

The driver went out with the changes described above and was able to make three laps before the check red flag fell at the end of the session.

On used tires he ran the quickest lap-time of the day .1, 51.27 and put his car on top of the chard P1.

When the driver got out of the car he said to the team engineer "that was an excellent move", the under-steer is gone and the front end settles right away after a bump "We have a very good car.

The engineer said "you can thank this gentlemen for this ", He came up with this solution.

What happens in this situation is that the un-sprung weight or mass of the front end is so much under dampened that it upsets the whole suspension assembly.

The tire loses contact with the road surface skipping over the small undulations instead of riding over it and loosing grip in the process allowing the suspension to use a longer travel in compression than necessary causing under-steer and increases tire wear.

The gas pressure of 18 bars produces a lift of 68.2 kg 150 pounds support or lift to the corner of the car

Increased compression damping and higher gas pressure results in less suspension travel,

With shorter suspension travel, the spring accumulates less energy and does not need the heavy rebound damping to control the kinetic energy of the suspension

27

Relation between a damper running in a dyno and performing in the suspension

A strut or damper performing in a car reacts to load inputs of the road surface and applies (when adjusted correctly) just enough damping to counteract the input forces, contrary to the dyno which moves the damper at a constant speed.

Any time the damping is adjusted to higher forces the suspension travel will decrease.

The most important part of the damping curve is the low velocity area, from 0 to 5 cm or 2 inch /second.

When the valving system is designed well the damper will apply so much damping that at low velocity that higher piston velocities are not reached.

When looking at a dyno sheet the higher velocity range is not so important because the suspension will not reach that high velocity range when properly dampened at low speed.

A damper running in a dyno shows the damping performance of the velocity of the dyno program.

It offers an insight in how a damping force builds up relative to the piston velocity both in rebound and compression.

It does not show how the damper performs when installed in the car or how it responds to the input of the road surface because the damper is moved around by the dyno without being able to slow down or dampen the velocity or the suspension.

There are ways to record suspension movements and store them in the computer but the problem is that with every damper, adjustment change, spring rate or anti rollbar change the suspension velocity and travel change also.

We are still confronted with the unknown factor of how much damping should be applied to make all suspension components work together in harmony.

I think that is a factor which should be recogknized and learned by racing engineers but to do that successfully they should have a better knowledge of how a damper works and what one can do to make intelligent changes improving the handling of a car.

When testing with an Indy car at mid-ohio, we started with the original spring / damper setup.

VEHICLE DYNAMICS AND DAMPING

The data showed a certain excursion of the suspension (dip) when running over bumps and road undulations; this was the best setup the team had found that worked on the track conditions at Mid-Ohio.

The lap times were not spectacular but it was the best they could do.

When we switched to the damper set we had prepared for the test and made the adjustments I had recommended, two things happened.

After running a couple of laps, the driver came back into the pits reporting that all under-steer he had experienced with the other dampers was gone and in fact the car turned in to well meaning that the grip at the front had improved a lot changing from understeer to turning in to quick.

We increased the compression damping front and rear and the car was neutral balanced cornering faster with faster corner speeds.

The data showed that the compression travel was half the excursion as before on the same track and the same road inputs which meant that the chassis could be lowered taking advantage of the better aero dynamics and provide a more stable platform moving through the air.

In our quest for more we lowered the chassis and increased the spring rates which again resulted in less compression travel but a better race car in cornering and normal running.

So using combination of the gas pressure, spring rates and compression damping as a tuning tool we were able to make the car quicker and handle much better.

The driver set the fastest lap of the complete test session that day.

Most engineers are relying too much on self-developed theories about how a damper should work in a suspension.

They do not have the in-dept. knowledge of damping, valving, hydraulics combined with gas pressure used in modern damping technology to be able to use it successfully in the setup of a race car but since it is one of the most important components on the modern race car.

A dyno runs a damper at a pre-set speed and the damper reacts to the velocity of the dyno, so what you see on a dyno sheet is damper performance at the velocity of the machine The damper is pulled and pushed around without having input on the performance.

When mounted in a suspension the damper restricts the suspension travel any time the damping is activated by suspension movement, the dyno graph just shows a different line.

So to get a real picture of damper performance in a suspension the dyno should react to damper adjustments (higher damping forces) by slowing down the velocity of the machine and restricting the travel, this would give a true picture of damper performance as it happens in the suspension of a race car.

Dynamometer design is a science which has to follow new insights and experiences gathered in the real world of racing and performance, the leading manufacturers are well aware of the fact that they have to keep up with te latest developments and insights in damper technology.

28

Working at the racetrack

Working with racing teams, it is always good to look ahead and prepare for the next race, incorporating improvements made during the last practice and race, write up a manual in which improvements in handling and setup are recorded.

I usually write a report listing the changes made during the last practice with lap time chart and race.

The report is send to team principal and engineer with instructions regarding damper adjustment and spring selection to make sure that we are keeping the line of development, addressing the points which can be improved and submitting a sheet with spring rate, damper settings anti roll bar settings, cambers etc.

It is a lot more difficult to write about car set up, spring rates and damper settings than working on it at the race track because when you are in the middle of the action and make a change, the results show immediately and corrections can be made if necessary but writing about it and leaving it up to the engineers implementing the changes you suggested is sometimes risky.

It is therefore always good to be with the team one or two days before practice starts to discuss and explain why certain amendments are made in the setup and come up with a plan to follow.

The car has been worked on and rebuild in the race shop, bump steer, camber, casters set to spec and the cross weights evened out but there are always some little items needing attention first.

The car needs to be base lined and the driver to get used to a different track.

As a consultant much of the success on the race track depend on the relationship with the team engineer, if He thinks that you threaten his position you can be pretty sure that the changes are not made according to the suggested setup.

It has happened several times that the engineer made changes to the setup completely contrary to the written advice with negative results.

Looking at the results it showed that the engineer was not following the suggested changes.

The role of a damper specialist is delicate, to be a good damper specialist you need to be a race engineer to keep the picture of the complete suspension package using your intuition in combination with technical facts.

You have to be able to communicate with the driver as a rce engineer keeping an oversight over the total set up.

My early education was automotive and later in the Dutch army, I had a technical function and it was there that I got my first encounter with racing.

I was stationed in The Hague working in a large maintenance department for passenger cars and light vehicles.

We happened to have an engine Dyno on which we ran tests on motorcycle engines, the army used Britisch made motorcycles.

By way of coincidence, there was a race team owned or managed by Mr Lex Beels who had racing cars equipped with Matchless motorcycle engines, one of the drivers was Mr Hutchinson a KLM captain and apparently well connected with the higher military brass.

After overhaul of the engines, they were dyno tested on the engine dyno in our department.

Later, when I was in charge of the Koni racing service department, Mr Hutchinson a friend of racing driver Ben Pon who was racing Porsches in Europe, was a regular visitor at most of the European races which we attended with our service team, knowing that we always had a couple of extra paddock passes for guests available.

Working in a R&D department of a major damper manufacturer is very educational, you learn a lot about damping and suspension in a hurry, because it is your job to make improvements and come up with new inventions, there are so many different applications and damper systems but I am fortunate to be able to concentrate on damper systems for racing and high performance applications and get drawn into racing services, which in the beginning consisted only of attending races but was quickly getting involved in the complete setup bringing damping forces in line with spring rates and anti-roll bar rates.

The indebt knowledge of gas/hydraulic damping systems in combination with the experience obtained with ride development at sports car manufacturers like Porsche and Ferrari is the best educative program one can go through.

We spend several weeks in a row at the Porsche or Ferrari factory and on race tracks driving and changing damper valvings, spring rates and antiroll bar settings, working on the ride and handling because the final results would be incorporated in the next model year.

When the program was finished, all settings were recorded in a protocol, they could not be changed till the next production year but before final approval, the cars were driven by company principals and engineering staff as a last test.

Some times we could not get the ride and handling right on one particular model car, it was a new model cabriolet, no matter what we changed the ride and handling remained below expectations.

The problem was not in the suspension but the monocock of the cabriolet design was to flexible, acting as a spring.

Adding some stiffness and more spot welds the problem was solved and we could make a nice riding car.

With experience we made a big step forward in understanding damping as an integrated part of the suspension, after having been through the development of a broad scala of applications culminating in the high level of knowledge we have obtained in today's work.

JAN ZUIJDIJK

Working at the race track with a new team causes always some apprehension, you don't know what to expect and sometimes have to go against the opinion of the team engineer, but as soon as the car is moving you get a better understanding what direction to follow

In 1994 I did a full season suspension development with the Trans-Am team of American Equipment Racing with Ron Fellows driving, when we went to the first test session at a little track close to Indianapolis.

Will Moody was the engineer; the car was set up in the traditional Trans Am way when we started working.

Will Moody is the kind of engineer who excepts all your insights and recommendations as long as you can explain to Him how it works and why it works, I enjoyed the chat sessions after each test day, when we evaluated the day's work with Ron and Will and I had to anwer all the questions about why we made the changes.

Once He is confident that you know your business, there are no arguments about changes.

I did not know Ron before we started working with the team, but I learned to know him as one of the best drivers in the field, always calm, friendly and highly professional with precise information about the handling of the car and very sensitive to changes.

I remember one test where the car had what Ron called, slides across the back under power, I made one adjustment to the rear dampers, turning the low compression adjuster up one click which closed an orifice of 0.4 millimetres.

He went out on the track and reported that the slides were gone, that is how sensitive a driver can be but also how a big car reacts to one small adjustment.

Someone before me had determined that the best way to set up a Trans AM car was "to hold up the car with the shocks and get traction with very low spring rates".

The problem is that dampers (shocks) without gas pressure do not hold up anything, a damper is a reactive device, reacting to suspension movement.

VEHICLE DYNAMICS AND DAMPING

They were using 350lbs/inch at the rear and 500 lbs. at the front of the car.

Well it might have sounded good but did not work at all, the car was not handling well with little grip or traction rolling too much and the lap times were as could be expected (slow.)

After a day running with the original set up, in the evening after the test, the dampers were exchanged for our JRZ system, causing the mechanics to exclaim that it would be a nice car for a Saturday night special at riverside or something to that effect.

When I had adjusted the dampers properly for the first outing, they could not believe the change it made in handling and running speed.

After the first couple of laps, the car seemed to run better, we could watch the car running through the turns and noticed the difference.

In addition to the damper adjustments we started working on the spring package increasing the spring rates to a respectable 800 lbs./inch front and 600 lbs./inch rear.

The car changed into a real race car, it looked well and stable through the turns, the driver was happy with the results and the best lap time was 3 seconds better than when we started. After we finished the test we went to the first race of the season in Miami.

In practice the car went very well but the team did not show the true potential in practice, it was in qualifying that car and driver out performed every other car on the track, He finally qualified with a couple of seconds faster than the next competitor.

It was an outstanding result; the only major changes were the spring rates combined with the JRZ dampers designed and build following my theory that a combination of gas pressure providing support to the suspension and controlling the unsprung weight with a double adjustable damper was a winning design.

Ron won most of the races with this setup during the rest of the season.

In the week before the race in Detroit, we tested on a little track in Michigan named Gratan.

This track was very bumpy at the end of the straight just before a 90 degree right turn, a perfect track to test the suspension set up.

We started the test with the setup of the last race.

If we would have followed popular believe, the first reaction would be to reduce the compression damping forces, lower spring rates and lower the gas pressure to reduce lift.

Well we did the opposite and to everyone's surprise, the car ran better and better with the higher damping forces, including through the rough section the heavier damping forces and higher spring rates did not bother the car at all, the car ran faster and had good traction and grip.

When I asked Ron whether he considered the handling to suffer from the heavier damping and spring rates, He said no, not at all, the car improves with every change we make.

As a matter of fact, when I worked with the stock cars in Brazil, we used to increase the gas pressure to 25 bars for qualifying, it provides more grip and heats the tires quicker for a one flyingqualifyning lap.

Of course the gas pressure was reduced to 20 bars for the race set up when the tires have to perform well for the duration of the race.

General handling conditions which can be changed or improved with damper adjustments.

Under-steer:

If the driver reports under steer, Increase the high speed bump one click at a time in Front till the driver reports better turn in.

When the low speed adjuster is in middle position, increase the low speed bump one click

Decrease low speed rebound damping one click in Front.

Increase gas pressure, the damper is designed to operate at 16 bars but a variation from 8 to 25 bars is normal and acceptable for operation under racing conditions depending on the race track and conditions.

Gas pressure can be checked with a special gauge showing the gas pressure in the reservoir without releasing gas.

The gas pressure should be checked and changed with the dampers extended so that the volume in the reservoir is maximal.

Increase the high speed rebound instead of selecting a stiffer roll bar

Once the dampers have been tuned to satisfaction, the roll bar can be adjusted to be in the middle position so that the driver has a choice to us a stiffer setting or go to a lower setting.

Recently working at the testdays for the Daytona 24 hours 2013, the car I was working with suffered understeer and excessive dive under braking, the solution was to increase the spring rate front and increasing the low pressure damping in front.

Over-steer

Over-steer or the car being loose can be a result of different conditions.

Low spring rates on the rear axle can cause the suspension to be compressed too much on one side causing the inside tire to loose traction.

This cannot be cured with damper settings or adjustments but the car needs heavier spring rates.

When this condition is present the over-steer will change in to under steer when the throttle is applied because the car will squat transferring weight to the rear causing the front wheels to get light.

After the spring rates have been adjusted, Increase high speed bump setting one click at a time at the rear till the driver reports better handling.

A second cause for over-steer can be too high spring rates rear and too high low speed rebound damping at the same time.

For persisting over steer, decrease low and high rebound damping and increase low speed bump damping at rear two clicks, Increase gas pressure with 5 bars.

Under racing conditions it is recommended to use about 20 bars (275) PSI when dampers are warmed up, 18 bars cold to prevent premature tire wear.

The 20 bar gas pressure does not cause a problem in itself but in combination with high spring rates and heavy compression damper settings it can cause the tires to heat up more than necessary.

With a softer spring package the gas pressure would not be as critical.

In general, tire wear depends for a great deal on the total stiffness of the suspension

This is the addition of spring rate, compression damping, gas pressure and tire pressure.

This is why it is important not to use higher spring rates, more damping, more gas pressure or roll bar settings than necessary to tune and make the car perform well.

The lift provided by the 20 bars of gas pressure is beneficial for the ride with the same spring rates.

A gas pressure of 20 bars produces a lifting force of 72 kg 160 pounds on the 22 millimeter piston rod, taking over part of the load of the car.

The spring has to carry less weight and can be lowered till the correct ride height is set.

The gas pressure carries part of the weight but does not add to the spring rate so the combination steel spring/gas provides a nicer ride without sacrificing handling.

29

Adjusting The Two Way Adjustable Damper

The two way adjustable gas hydraulic damper system, is equipped with an external reservoir.

This is in our experience the best option to control (dampen) suspension movements and dynamic forces of a high performance vehicle in motion, especially the un-sprung mass benefits from the large displacement piston rod.

The external reservoir is necessary to accept the fluid volume displaced out of the working cylinder without increasing the gas pressure

Using an external reservoir ads an extra function to the operation of the damping system on the compression (bump) side of the damper.

The adjustment of a two way adjustable damper in a race car suspension is the most basic but very effective damping system of the external adjustable damping systems.

Rebound Adjustment

The rebound adjuster is operated by a small diameter shaft running through the hollow piston rod. controlling the flow of fluid through orifices and adjuster nut closing or opening orifices with each click of adjustment.

Each time an orifice is closed the nose angle of the damper changes to a more upright position.

The adjustment system operating with a certain number of small orifices gives better results in damping adjustment performance; a drilled orifice is more accurate and can be reproduced with a high level of accuracy because the flow through the orifice is always the same.

A high level of accuracy is important for the damping forces to remain at the same values between different dampers.

When the flow through the orifices is restricted by the adjuster a larger amount of fluid is diverted through the valving system causing higher damping forces at low and high piston velocity.

There are systems working with one big orifice, closed gradually by a tapered needle, when the needle is turned deeper into the orifice, the flow of fluid is restricted and a larger amount is forced to pass through the piston valving.

When the adjustability of the damper is used to tune the handling of a car, the only choice is adjusting either rebound or compression forces.

In addition to adjusting the damping forces, the gas pressure can be used to tune the handling of the car as well.

The compression or bump damping forces are adjusted by the valving located in the top of the external reservoir.

In the two way adjustable damper the high damping forces are adjusted, every time the bump adjuster is turned one click, the damping force reaches a higher level of adjustment, showing up on the graph in nice even increments.

There are no orifices in the compression valving of the two ways adjustable damper, the high velocity damping is more effective since there is no fluid flowing through orifices.

Assuming that the spring package and antiroll bar settings are within the required ballpark, the damping forces need to be adjusted to control the sprung mass, suspension travel and keeping the tire contact patch firmly in contact with the road surface.

If the damper is equipped with a valve to fill gas, the pressure should be set at a reasonable starting value which is about 16 bars or 225 psi.

It is always good to start with the compression damping forces on the firm side in order to make the driver feel the different damper settings.

The damping forces of the JRZ adjustable dampers are designed to be used in about the middle of the adjustment range so that damping forces can be adjusted up or down when adjustments have to be made.

Starting with a setup in which suspension components like spring rates, anti-roll bar rate are on the strong (high) side makes it more difficult to differentiate between the performance of one component or the other and find your way through the adjustments.

For a starting position I would set the rebound at position 12 and the compression adjuster at 7 which is right in the middle of the adjustment range.

There are 24 positions in rebound and 14 in compression providing finer adjustment steps.

The gas pressure should be at 16 bars for a start.

Running a few laps should give a good indication what direction to take with the damper suspension tuning.

For instance when the driver reports the car to be too soft that is a good sign because now we know that we can improve on the handling by increasing damping.

Start increasing the bump or compression damping first, this will have an immediate improvement in traction and handling because it slows down the bump travel and velocity of the un-sprung weight of the suspension assembly.

Remember a two way adjustable damping system adjusts the low and high velocity damping force at the same time. (Simultaneous).

When the driver is not satisfied with the car and reports it still to be on the soft side, increase both rebound and bump one click front and rear and repeat this till the car starts to get a little nervous, now it is time to stop increasing the rebound damping forces because adding more damping will make the car too roll stiff and it may start to slide and loose grip when cornering

When the compression damping has to be adjusted to the maximum position, I would increase the spring rates, it is a sign that the spring package is on the soft and does not support the sprung weight as it should be.

To low spring rates cannot be compensated with heavier damping forces, the spring package needs to be in line with the sprung weight and down force.

When the spring package is updated, one can try two different adjustments, first increase the compression damping with one click and if the under / over steer condition persist, reduce the rebound damping one click.

This is to make the driver feel that the car is better in contact with the road surface, every time the compression damping is increased, the car will gain grip and traction.

The dampers have to be adjusted front and rear with the same number of clicks to keep the balance front to rear.

When the handling of the car gets better and the laptimes faster, the driver might get closer to the edge and starts experiencing under or over steer conditions.

Over-steer or (being loose) can be cured by increasing the bump damping at the rear one click at the time each run.

One click adjustment does not seem much, but the car reacts directly to low velocity changes.

When the car gains grip and traction at the rear, the driver might experience the change as under steer or push, with better traction at the rear, the front end of the car might be lacking grip compared to the rear and the car starts under steering.

Now the front bump damping has to be increased one click in order to balance the handling front to rear.

Because the damper responds immediately to the adjustment at a very low piston velocity, the car reacts to each click of damper adjustment so the driver will feel each click as a change and in most cases, a positive change.

Doing it this way the handling will improve alternately front and rear till a well handling and balanced car is achieved.

The 16 bar or 215 psi pressure provides 60 kg or 134 lbs. of lift to the car and helps carry the sprung weight, planting the tire better against the road surface.

Increasing the gas pressure up to a maximum value of 20 bar or 275 psi will result in better grip and traction.

Increase the pressure with 5 bar front and rear, send the car out and watch the result on the stopwatch and hear the comments of the driver, this adjustment is almost fail proof, there might be circumstances in which the settings of other suspension components prevent improvement in handling, but increasing the gas pressure is always an improvement in grip and traction.

With all other suspension components tuned in harmony the gas pressure can be used up to 20 bar, 275 psi without a problem.

Setting up the car this way I am sure that the results will be better than expected, one click adjustment can make the difference in handling.

ADJUSTING THE MONO TUBE SINGLE ADJUSTABLE DAMPER

The single tube or mono-tube gas hydraulic damper needs 25 bar or 340 psi gas pressure in order to operate and reach high compression damping forces

When the gas pressure is lower or non-existent, the damper cannot work, there are miss understandings about the function of a gas volume under pressure in the mono-tube damper, it is not a magic medium but necessary to make the system work.

The single tube gas damper can be adjusted in rebound only because in most designs orifices work in rebound and compression simultaneous but when the rebound is adjusted, the bump force will increase with a certain but small percentage of the rebound value.

There is nothing wrong when the bump force increases a little with the rebound force, we have had questions about that, some manufacturers build in a small non return valve to prevent the bump force from changing but the driver does not notice the difference, as a matter of fact, it will have a positive effect on the handling

I prefer to work with small, drilled orifices which are closed of one by one by an adjusternut or sliding valve, drilled orifices are more precise.

When the rebound forces are adjusted, the compression force will react too, because the forces are adjusted by closing off the free flow through orifices, reducing the flow with each adjustment.

Both rebound and bump forces react to adjustments simultaneous, the compression force reacts to a much lesser extent than the rebound damping force, because the piston area responsible for the rebound force is larger than the piston rod area responsible for the bump forces.

It is this ratio between the two areas's which determine how much the compression damping force will increase with the rebound adjustment.

The nose angle of the graph increases to a steeper angle while the high velocity damping increases simultaneous.

The single tube damper needs to operate with a gas pressure of 25 bar, 350 lbs., this is necessary for this damping system to reach the max compression or bump forces 25 bar of gas pressure delivers a back pressure, supporting the fluid Colom, with 375 kg or 825 pounds. 25 bar is the pressure necessary to reach the highest compression damping forces, not to be confused with the lifting force because the lifting force is the pressure of 25 bar times the cross section of the piston rod which is in most cases 12 mm or one square centimetre so the lifting force is 25 kg or 55 pounds.

30

Single Tube Damper With External Reservoir

In a single tube damper system with external reservoir, the compression force does not rely on the presence of 25 bar, 350 lbs of gas pressure, the gas pressure can be variable and be as low as 5 bar, 70 lbs because the compression forces are generated by the resistance fluid encounters flowing through the valving system first before reaching the divider piston in the reservoir.

The compression valving is located in the top adjuster housing of the external reservoir.

The gas pressure in a single tube design with external reservoir ads a new function to suspension control, the gas pressure working against the cross section of the piston rod provides lift to the suspension of a corner of the car.

The amount of lift depends on the gas pressure X the cross-section of the piston rod.

This system has an advantage over the single tube hydraulic gas damper, the twin tube hydraulic gas damper and the twin tube hydraulic damper with atmospheric pressure.

We prefer larger fluid displacement per centimetre stroke, the piston rod diameter can be as much as 22 to 25 mm.

Larger diameter piston rods provide more lift and better handling.

Damper adjustments have a great influence on controlling the vehicle dynamics of a moving race or sports car.

The rebound damping force is responsible for controlling the sprung weight (mass) of the vehicle.

The compression damping force is responsible for controlling the unsprung weight (mass) of the vehicle.

The gas pressure preloads the suspension, pressing the tire contact patch against the road surface.

The damping control at low piston velocities keeps the platform from making big excursions helping the aero dynamic devices on the car have a better effect.

A vehicle at speed is subject to several load inputs into the suspension, those force inputs are not constant but are intermittent, and vary with every inch a car is driven over a certain road surface, in straight line or cornering.

The force inputs centre on the ride height of the car moving in compression and rebound alternately, crossing the ride height line with the frequency of the sprung weight and un-sprung weight.

The most problematic conditions for a race driver on a race track are:

Under-steer condition or the car is pushing, the front end of the car continues in a straight line when the steering wheel is turned, which takes considerable time before a car with an under steer condition can accelerate again out of a corner,

Over steer condition or the car is loose, the car breaks out at the rear because of low grip levels, the car slides out of the corner with all four wheels.

Poor traction, lots of wheel spin when putting the power down accelerating of the corner

The handling of the car being upset when running over every bump in the road or running over curbs.

Over or under steer condition is a result of the roll stiffness of the front or rear-end of the car.

VEHICLE DYNAMICS AND DAMPING

It can be that the spring rates are too high in relation to the sprung mass of the car, a good way to find out is to calculate the natural frequency, and the natural frequency of the sprung mass is determined by the spring rate and weight or mass of the car.

1:2pi X sq root out of spring rate: weight of a corner of the car.

The natural frequency can be used to select a certain spring rate.

The antiroll bar is to stiff or the antiroll bar settings to high.

Too much low velocity rebound damping preventing lateral weight transfer in combination with too much high velocity rebound damping.

Too roll stiff a front end will produce under steer or push, the front wheels will not follow the direction they are turned in but slide to the outside of the turn.

The same condition at the rear of a car will produce over steer.

The rear tires will not roll through a turn but slide to the outside of the turn producing wheel spin under acceleration.

The roll stiffness of the frontend or rear-end of a car is an accumulation of spring rate, antiroll bar settings, damping force and tire pressures and can be changed by adjusting either one of the components but it is necessary to find out which one is the overriding factor.

If the sum of all those settings is too high the car will under or over steer or both.

It is remarkable that race engineers almost always tries to find a cure for bad handling, under steer or over steer, by increasing the rebound damping while there are so many different ways to arrive at a good handling and balanced car.

Many do not recognise the importance and benefits of the direct acting compression damping combined with lift provided by the gas pressure.

But when they do, they all report much better handling and laptimes.

Only a couple of weeks ago, a race team with two BMW's replaced the original dampers with a complete set of JRZ Spring/Damper Suspension units.

Without changing anything else but the suspension well tuned by our engineer, the cars ran 2 seconds faster laptimes, qualified on the front row and won both races.

The announcer unaware of any technical changes called it a surprise win of the team cars.

When the damping force is increased on the rebound side of the curve, the extension stroke of the suspension will be slowed down to quickly, and the handling will be worse than before the adjustment was made.

The reason is that the rebound damping force, special at low velocity resists the extension of the suspension, slowing down the extension rate of the suspension means that the spring is not able to restore ride height of the vehicle before the next bump compresses the suspension again.

The spring, wants to restore ride height after compression, giving back the energy accumulated in the compression stroke but the suspension not being able to restore ride height quick enough results in a dynamic lower ride height of the car.

People will say, that is the jacking effect as if that is something unavoidable but at the same time, the suspension, will pick up the tire of the pavement reducing grip.

The jacking effect is not something we have to accept as a fact of life it has been dialled in and can be dialled out again taking the right steps a car without the jacking down effect will be a better handling car and responds much better to setup changes and road conditions.

When the suspension of a car jacks down by the motion of the car, it is pure a matter of too much low velocity rebound damping.

If the low velocity rebound damping is too strong for the spring force the suspension will start the extension stroke but in the process is picking up the un-sprung mass, wheel, and brake, control arms etc. takes the complete suspension assembly with it on the rebound stroke and takes the load of the tire.

The worst part is that the tire contact patch is losing contact with the pavement and in fact loses traction and grip.

VEHICLE DYNAMICS AND DAMPING

Under those conditions the tire will skip over the road surface at a fairly high frequency, the natural frequency of a tire is around 10 to 15 hertz or 900 cycles per minute, so the un-sprung mass of a vehicle is actuated in that same frequency.

Especially with the current tendency to keep the compression damping as low as possible and dial in a high amount of rebound damping, the suspension assembly will try to move in that same frequency as the tire which is a cause for bad handling and low grip levels.

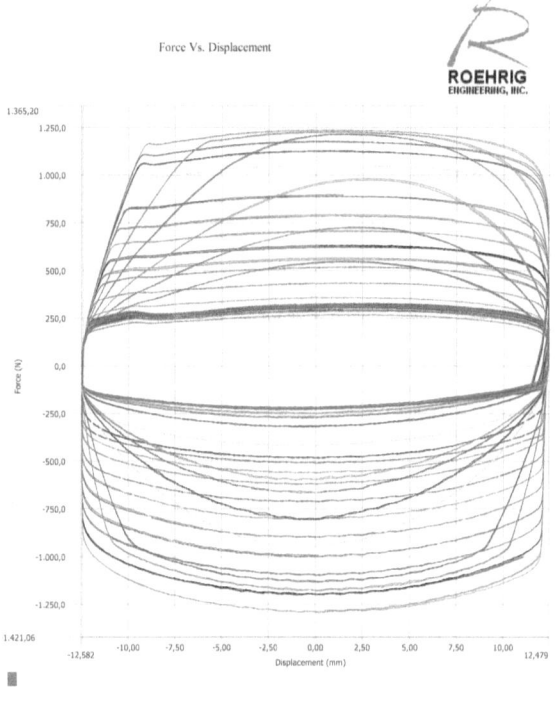

Compression and rebound adjustments

Since most engineers believe in using the rebound damping to correct handling problems, they don't get optimal use out of the suspension setup and get stuck with a car not responding to adjustments of other suspension components.

The car will respond by being nervous and might be difficult to drive around corners.

Also acceleration and under braking the car will have low grip levels resulting in wheel spin under acceleration and locking up the wheels under hard braking.

Another adjustment tool is the variable lift the gas pressure working on the piston rod cross section is offering.

The lifting force of the gas pressure can be varied from 5 bar to 25 bar giving a lifting force with a 22mm piston rod of 18,95 KG at 5 bar to 94,75 Kg at 25 bar.

Maybe the unknown territory of working with gas as a suspension medium and setup tool is cause for hesitation, the first reaction is to lower the gas pressure.

The gas force is a lifting force aiding the coil spring carrying the sprung mass of the car without increasing the spring rate.

At the same time the gas pressure preloads the suspension with an equal pressure over the travel of the suspension.

This is a very important observation because there is a fundamental difference between changing springs with a higher spring rate, or increasing lift to the existing lower spring rate using gas pressure.

The spring rate graph below shows how the gas force ads lift to the suspension without increasing the spring rate.

The original spring rate is the second line from the top, the spring rate is 50 kg/centimeter, 275 lbs/inch, second

The top line shows the spring rate with the gas force adding about 100 pounds lift to the spring force over the total length of the spring rate line.

The spring force to carry the car at ride height is 390 kg, 860 pounds.

Lowering the spring rate from 50 to 45 kg or 250 lbs./inch and adding the gas pressure, the spring force combined with the gas pressure force delivers exactly the same force at ride height to carry the car/sprung weight as the spring did without the gas force

This is very important to know because the gas pressure is adjustable from 5 bars to 25 bar and at 25 bars, 350 lbs./inch the lift would be 95 kg or 208 pounds.

The lifting force is present over the entire length of the stroke without increasing the spring rate.

This means that for the same sprung weight, lower spring rates can be used while the gas pressure carries a substantial part of the sprung weight of the car.

Being able to use lower spring rates means that the natural frequency is lower also, a lower natural frequency translates into better traction and mechanical grip.

Every time we apply this principle to a race car, it is amazing how the car reacts with better grip, power down and acceleration

Also the cornering improves a lot, accepting higher G forces and a smooth ride over the bumpy parts of the track.

When installing a spring with a higher spring rate, the spring reacts to compression of the spring; it will react with the same force as the force needed to compress the spring.

For example if we take a spring with a spring rate of 250 lbs./inch it means that the spring needs to be compressed over a length of 1 inch with a force of 250 lbs.in order to deliver this force of 250 lbs. back.

Using a spring rate of 300 lbs/inch will result in a higher natural frequency

To avoid the higher natural frequency, we can combine the 250 lbs/inch spring with a 20 bar gas pressure and have the same carrying capacity with a lower natural frequency as before.

Gas pressure working against the cross section of the piston rod is a constant force which does not depend on deflection or compression of the suspension, it is always there even when the car is not in motion, the gas force is always present at every part of the stroke whether it is compressed, extended or at ride height.

This is a result of calculating the right gas volume for the displacement of the piston rod and the stroke of the damper.

Another positive aspect of using a combination of gas pressure and damping is that no bump rubbers are required to prevent the suspension from bottoming out.

31

USING BUMPRUBBERS

Bump rubbers only work when the suspension is compressed far enough to touch and compress the bump rubber

When the use of bumprubbers is necessary to support the suspension there is something wrong with the spring rates (to soft) or the rebound damping to strong.

When a suspension touches the bump rubbers on impact, it implicates that the compression stroke of the suspension is not supported by the springs and probably too much low velocity rebound damping is dialled in jacking the car down.

The sudden increase in spring rate at the point where the bump rubber is compressed upsets the car's handling profoundly.

The suspension will react with a sudden upswing of the carbody when the suspension stroke is reversed in rebound and much rebound damping is applied.

Some people like to use packers on top of the bump rubber moving the point of contact (impact) closer to the extended position of the suspension.

Packers are solid shims of aluminium or nylon which are split and held together by wire so that they can easily be slipped inside the coil spring and change the point of contact/impact with the bump rubber closer to the ride height position of the vehicle.

In some cases people go overboard adding so many packers that the car is riding on packers and bump rubbers defying the purpose of the suspension.

Of course when a car needs that many packers in order to keep it of the road it would be better to increase the spring rate, it is clear that the spring in this package does not provide enough lift to carry the sprung weight in addition to the dynamic force.

The use of packers absolutely contradicts the purpose of suspension design.

A suspension is a flexible devise designed to absorb energy of motion over a certain suspension travel, the length of the travel determines how much kinetic energy is dissipated.

The absorption of energy needs to take place over a certain stroke or travel dampened by the compression damping of the damper.

Using a bump rubber with packers changes the spring rate abruptly to a much higher rate at point of contact causing a high impact force; at this point the kinetic energy will not be absorbed over a certain length of travel but over a very short part of the travel.

That is exactly the situation we like to avoid, as a result the travel over which the energy is absorbed is to short and the sprung weight X the acceleration is violently stopped, running out of room (stroke).

This is felt as a violent force which will pitch the car violently upward and when the rebound settings are at a high rate, the suspension will not extend itself quick enough so that the tire leaves the road surface temporarily.

The result is that when the sprung mass (carbody) changes direction into the compression stroke, the damper has not extended itself far enough in time to absorb the energy over the length of the stroke needed to do this effectively.

32

Adjusting The Rebound Damping Force Of The Threeway Adjustable Damper

In a three-way adjustable damper the high and low velocity rebound forces are adjusted simultaneously.

This is because most damper systems use the orifices or orifice to adjust rebound damping restricting the free flow of fluid through the orifices, although there are damper designs adjusting damping forces by increasing the preload on the shim stack or blow-off valve.

In most designs the rebound adjuster shaft runs through the hollow piston rod operating a tapered needle or an adjusting nut closing of orifices one by one with every click or part turn of the adjuster.

When closing orifices thus restricting the flow of fluid from the upper part of the cylinder to the lower part of the cylinder which is the low velocity damping, the nose angle of the graph shows a steeper angle but at the same time the flow of fluid through the valving increases so that the high and low velocity damping increase simultaneously with every click or turn of the adjuster.

The high velocity rebound increase in even increments when the rebound adjuster is turned from the zero position to the maximum position,

The compression high and low velocity damping forces are adjusted separately from each other

The low velocity damping is adjusted through closing off precisely measured orifices or with a tapered needle closing of one larger orifice.

The high velocity compression damping forces are adjusted, increasing the preload on the valve stack so that the fluid in the working cylinder builds up to a higher pressure each time the adjuster is turned one position or click from zero to the maximum.

The damper graph shows this increase in damping force in nice even increments, each click representing a new position.

The blow-off valve design is very digressive so that the damping force remains at an even level over the length of the stroke.

33

Adjusting The Fourway Adjustable Damper

What we call the four way adjustable damper is actually a five way adjustable damper.

In addition to adjusting rebound and bump damping forces in low and high velocity, the adjustable gas pressure is a fifth adjustment and a very effective adjustment tool, so in reality it is a five way adjustable damper, the gas pressure can be varied from a low of 6 bar 85 lbs. to a high of 25 bar, 350 lbs.

In the four way adjustable gas / hydraulic damper both the high and low damping forces in rebound as well in compression are adjusted independent from each other.

This design allows the damping forces to be adjusted in a perfect symmetrical way.

In rebound as well as in compression, the damping graph shows the nose angle and the increase with each click of adjustment

With this design any possible configuration of damping graph can be achieved just using the adjusters.

In the sixties and seventies dampers had to be re-valved for every race track and every race, this is not necessary anymore since the damper is adjustable 5 ways and race suspensions have gone through a long process of development.

The nose angle of rebound and compression can be altered independent from each other in the low velocity damping.

The high velocity of rebound and compression damping can be changed using the adjusters with a very large adjustment range.

In addition, the gas pressure providing lift supporting the sprung weight with a variable lift from 10 kg to 95 kg depending on the gas pressure.

Adjusting the damping forces at the race track is a little more complicated because of the choices available when adjusting the damping, but it is also more rewarding because of the available mix of damper inputs.

Damping forces have to be applied in response to the handling characteristic of the car.

The question is what part of the damping force graph to adjust when the car comes into the pits and the driver comments on the handling.

Keep increasing the compression damping one click at the time.

When the low velocity damping is adjusted to firm, the car will react by lifting the tire of the road surface and the car will lose traction and grip.

The tire lifts of the road in high frequency but small amplitude, enough to loose grip and traction.

The high velocity rebound force is not so sensitive because the damping force acts at a velocity of 5 to 10 centimetre per second, (2 to 4 inch per second.

When adjusting the dampers at the race track, I would start with low rebound settings, low velocity damping at 2 clicks and high rebound settings at 2 clicks, front and rear.

The low velocity compression damping set at position 3 and the high velocity compression damping set at 6 or 8 the gas pressure at 16 bar front and rear.

Leave the antiroll bar at a low setting, because when the antiroll bar is to strong the car will not react in a positive way to damper settings and spring changes, the high antiroll bar settings will overpower the other components like spring rate and damper adjustments.

This is a good starting point and will give the engineer a good indication what direction to take after the car has been base lined.

From this point forward the handling should be improved by adjusting either the damper settings, gas pressure or spring rates, increasing the low speed compression damping one click at a time.

When the car responds in a positive way, increase the high speed compression damping one click at a time till the suspension becomes to stiff.

Back off one click on both low and high speed compression adjusters.

Now start increasing the rebound low speed adjuster one click at a time till the car is to rigid and back off a couple of clicks, after this increase the high speed rebound damping one click at a time till the car reacts with less lateral grip.

Back off 2 clicks and run the car with this setup, check laptimes and driver comments.

Keep in mind that any change made to the suspension is directed at keeping that small tire contact patch in firm contact with the road surface.

If any suspension adjustment does not result in better traction, grip, cornering or make it worse the engineer should reconsider whether the adjustment should stay or put back in the former position.

It is easy to make the wrong choice in adjustment and when not corrected immediately losing the positive direction, so the best way for a successful setup is to be very methodical, go by the facts, not by imaginary solutions, because a race car responds to laws of physics and logic, not to something we think it should do.

The best theoretical set up is useless if the race engineer is not able to deal with the dynamic situation at the race track, when the car runs at speed and the car reacts different than expected.

In a recent exchange with a race engineer, the car was set up properly, using all the data available excluding all possible surprises.

At the race track however when the car ran at high speed, the driver experienced a notorious under steer at high speed which could not be cured in time.

Although the driver was able to put the car on the pole, the high speed under steer condition caused so much wear on the left front tire that the tread was shredded and the team could not make the race, packed up and went home.

Normally the tire temperature and the condition of the tire tread surface indicates how severe the under steer can be and certainly the driver can explain where on the track the condition manifests itself.

Looking at the data, after the fact, it occurred to me that there could have been a couple of changes made to alleviate the under steer problem.

First the high speed under steer condition points to a possible lack of down force at the front compared with the rear-end of the car.

Secondly, I would take a look at the front rebound damper settings, if they are too high, turn them down because high damping forces because a dynamic roll stiff condition of the front axle.

In addition check the anti-roll bar settings of the front axle, when they are too heavy, the antiroll bar in combination with the rebound settings can cause severe under steer conditions.

The spring package showed a significant difference in spring rates front to rear with the rear spring rates significantly higher than the front which can cause an imbalance in handling front to rear, although the rear motion ratio was a little higher.

In this case it seemed that the rear tires had more grip and the car unloaded the front tires more under power than the car set up could handle.

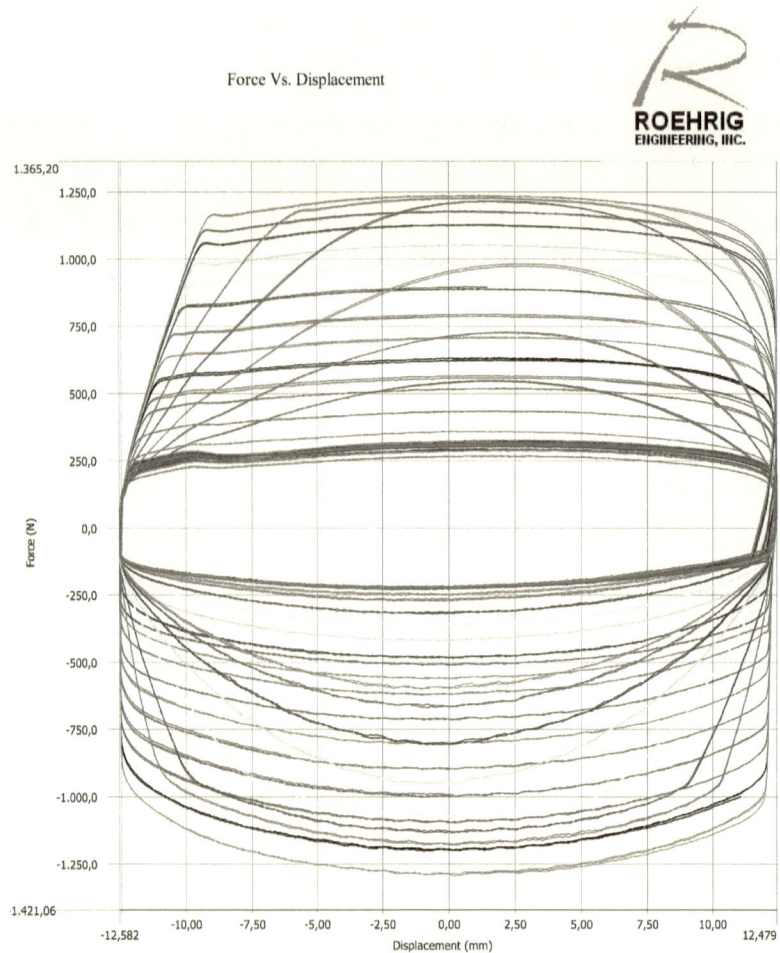

Compression and rebound adjustments

34

Force Velocity Graph Versus Force Displacement Graph

When comparing a force velocity graph with the force displacement graph of the same damper run in the same dyno with the same adjustments, there can be a distinct difference between the two graphs.

The force displacement graph shows the damping forces to start increasing right from the beginning of the stroke at very low velocity.

Whit the low velocity adjuster in the maximum position, the maximum damping force is reached within the first 7 millimetre of the stroke.

The force velocity graph recorded using exactly the same damper, run in the same dyno with the same adjustment, shows the damping force to increase, reaching maximum damping force after 25 millimetre of the stroke.

There seems to be discrepancies in damping performance between the two methods of testing the damper, while the dyno and the damper are the same as before.

Well there is nothing wrong with the damper, the damper performance does not change using a different testing method, the difference is in the way the damping forces are measured.

When taking a force velocity graph, the machine runs at the same speed (piston velocity) while the damper is adjusted through the adjustment positions Each adjustment position is recorded and shows a different line right from the start of the stroke with the damping force increasing directly from the start of the damping cycle.

The force velocity graph is run at the same stroke of 50 millimetre of 2 inch but the dyno starts at a slow velocity with increasing till the maximum test velocity has been reached.

The dyno measures the damping forces right in the middle of the stroke so the maximum force at the lowest velocity does not start at the beginning of the stroke but in the middle of the stroke at 25 millimetre or one inch into the stroke.

To produce a correct force velocity graph the machine should run different strokes, starting with a stroke of 10 millimetre, 20 millimetre, 30 millimetre 40 millimetre and 50 millimetre.

Mixing this with different rpm of the dyno one can produce a true force velocity graph of the damper.

Using different strokes and diffent rpm, the dyno records 5 different piston velocities or more, this way the damping forces are recorded in the middle of the stroke and the damping force increase is in line with reality.

35

Back ground Jan Zuijdijk

Jan Zuijdijk has been active in motorsports as suspension engineer and designer of damping systems for over 45 years, specializing in gas hydraulic damping and vehicle dynamics as applied to race cars in a diversified field.

His career started at Koni in Holland in 1960 in Research and development and was soon a regular suspension engineer on the Formula One and international racing series providing set up advice to the teams.

Jan Zuijdijk together with the late Henk Richten, service manager at Koni Holland, designed and pioneered a new revolutionair damper design of which the damping forces could be adjusted without taking the damper of the car.

After the new design damper was tested on the Ferrari prototype sports cars at the Nurburgring, the Formula One teams of Lotus, Tyrrell, Brabham, Surtees and other teams started using the new damper designed by Zuijdijk and Richten.

Zuijdijk worked within the research department on a Koni Gas hydraulic Self Leveling Suspension system for the Porsche 911 and Ferrari 365 2+2 especial tuning the ride and handling of the system.

The Leveling system using suspension movement to activate the pump action providing lift to carry the variable weight of the car.

In 1972, Jan Zuijdijk moved to the United States of America working with the Koni importer in the function of Chief Engineer and became directly involved with the American Racing series of which there are many and diverse series, soon the Zuijdijk prepared Koni dampers dominated the Indy car series.

In 1978, Zuijdijk accepted the function of Director of Engineering and Racing services, responsible for product development for the North American and Canadian market, organizing racing service throughout the country at Koni America in Culpeper V.A.

In 1988, Zuijdijk resigned from Koni and continued to work as consultant to racing teams in the USA.

In 1989, He worked with the Patrick racing team, designing suspension components for the Indy car team, working closely with Emerson Fittipaldi on the car setup, winning most races and the championship.

In 1991 Zuijdijk returned to Europe, starting the JRZ Suspension Engineering business, to develop suspension products designed for the demands of modern race car suspensions and vehicle dynamics.

The objective was to take the experience with the different damper designs and combine the features In a successful design

Zuijdijk designed a hybrid system using a combination of hydraulic damping and gas force to provide extra lift for the suspension of race and road cars which is very successful in many applications all over the world.

In 1994, Zuijdijk worked on a chassis development project with General Motors winning most of the races in that season with driver Ron Fellows and Will Moody engineering.

The JRZ damper system is now a race proven design having won major races all over the world and is a major player in the high performance suspension industry worldwide.

Zuijdijk is still active in damper development, consulting, seminars and in race car engineering. jzuijdijk@hotmail.com

www.ingramcontent.com/pod-product-compliance
Lightning Source LLC
Chambersburg PA
CBHW030922180526
45163CB00002B/434